At the End of the Shift

Mines and Single-Industry Towns in Northern Ontario

Edited by Matt Bray
Ashley Thomson

16
EasyRead Large

RHYW

Copyright Page from the Original Book

Editor: Paul Litt
Copy editor: Nadine Stoikoff
Design and Production: GSN
Printing and Binding: Gagné Printing Ltd., Louiseville, Quebec, Canada

The editors acknowledge with gratitude the funding assistance of the **Social Sciences and Humanities Research Council** both for the 1990 conference and for help in preparing this manuscript for publication.

Dundurn Press wishes to acknowledge the generous assistance and ongoing support of **The Canada Council, The Book Publishing Industry Development Programme** of the **Department of Communications, The Ontario Arts Council, The Ontario Heritage Foundation** and the **Ontario Publishing Centre** of the **Ministry of Culture and Communications.**

Care has been taken to trace the ownership of copyright material used in the text, including the illustrations. The authors and publisher welcome any information enabling them to rectify any reference or credit in subsequent editions.

J. Kirk Howard, Publisher

Canadian Cataloguing in Publication Data

At the end of the shift: mines and single-industry towns in Northern Ontario.

Papers presented at a conference held by Laurentian University's Institute of Northern Ontario Research and Development, October, 1990. Includes bibliographical references.

ISBN 1-55002-150-8

1. Mineral industries–Ontario, Northern–History. 2. Cities and towns–Ontario, Northern–History. I. Bray, Robert Matthew. II. Thomson, Ashley. III. Laurentian University of Sudbury. Institute of Northern Ontario Research and Development.

HD9539.A3C35 1992 338.2'09713'1 C91-095608-1

Dundurn Press Limited
2181 Queen Street East
Suite 301
Toronto, Canada
M4E 1E5

Dundurn Distribution Limited
73 Lime Walk
Headington, Oxford
England
OX3 7AD

TABLE OF CONTENTS

Tables, Maps and Figures

Comme je I'ai connue

Ce n'est point une ville
Trente cabanes de planches
Et des maisons mobiles semées dans le paysage
Solitude savaneuse
Autour d'une tête de puits.

Le commerce des biens et des corps
Se passe à la Blind ou dans le shack de la Quirke
Marie tient bordel
Au fond du pit de sable
Et Georges y vend du blanc
Y saigne le Portugais, l'Espagnol et Frenchie
Peut-être l'Allemand
Fraichement évadé de l'holocauste.

Au fond du puits
On parle le langage des signes
Mais le bras arraché ou la jambe écrasée
Se crie sonore
Que personne ne comprend.

Le soir dans les baraques
On couche à deux ou trois
À cause des femmes absentes
La partie de cartes du vendredi
Se termine lundi à l'aube
Paie flambee de deux semaines
L'autre s'élance dès la sirène

Sur le sentier minimal de terre
Serpent de poussière liquéfiée
Qui tousse la bagnole de Rhéal
Et des monstres éreintés.

Au bas de la pente près du lac
Le Boss habite protégé equipe
Premier Maître
Maitre des contres et des esclaves à gages
II vient des USA
Percer le coeur du Boucher
Pour que la bombe éclaire de nouveaux Hiroshima...

Pascal Sabourin, professor of French literature at Laurentian University, reflects on his experiences as a student in 1956-57 in the mines at Elliot Lake.

INTRODUCTION

Matt Bray and Ashley Thomson

MINING AND THE COMMUNITY IN NORTHERN ONTARIO

Mining has obviously played a formative role in the modern history of Northern Ontario. Since the mid-nineteenth century, along with forestry, it has been one of the key generators of wealth in the area. Because of its labour-intensive and stationary nature, moreover, mining has also been responsible for much of the urban development in Ontario's northland. The "cut-and-run" methods of the early forest industry required few permanent settlements; indeed, Northern Ontario is dotted with ghost towns such as Michael's Bay on Manitoulin Island, communities that for a few short decades served the local timber operations and then with them swiftly disappeared.

Mining was something else again. Nearly every large community and many smaller ones in Northern Ontario owe some measure of their existence to the industry. For some, such as Cobalt, Timmins, Kirkland Lake and Elliot Lake, to name only a few of the largest, mining has been virtually the sole *raison d'etre.* For places like Sudbury, it has been the single most important economic activity, while it has made significant though less direct contributions to the local economies of centres like Sault Ste. Marie and Thunder Bay. Remove mining from the North and its urban development would certainly have been less extensive and radically different in character.

Since the first shaft-sinking at Bruce Mines in the late 1840s, the fortunes of the mining communities in Northern Ontario, like the industry itself, have oscillated greatly. Largely single-industry based and dependent upon externally controlled companies, uncertain ore bodies, and wildly fluctuating, mostly foreign, markets, they have experienced the quintessential "boom and bust" existence. Eras of rapid economic and population growth have been punctuated by sudden collapses, frequently of devastating severity, characterized by high unemployment, out-migration and general business failure. A modern but altogether too typical case is Elliot Lake. Since its founding in the mid-1950s it has undergone no less than three such cycles, although the latest downturn is potentially the worst.

Unpredictability has been the only certainty for Northern Ontario's mining communities. At times, as in the immediate post-World War I period, they have accompanied the rest of Canada into economic recession. At other times, they, or at least some of the more fortunate among them, have trodden a different path. During the 1930s, for example, the growing demand for gold and nickel ensured that places like Timmins and Sudbury were spared the worst ravages of the Great Depression. Conversely, the mining communities of Northern Ontario have not always shared in the country's prosperous times. The removal of gold from the minerals priority list by the federal government early in World War II struck Timmins and Kirkland Lake a bitter blow that continued to hurt long after the war had ended.

Since World War II, in particular, this uncertainty has prompted suffering communities to look to senior levels of government for the economic stability that has not been provided by free enterprise. Although expectations of assistance from federal and provincial authorities have increased exponentially across all of Canada in the last half-century, the persistence of economic difficulties in the mining sector has made this option ever more popular in Northern Ontario.

This trend was perhaps best symbolized at the provincial level with the creation of the Ministry of Northern Affairs by the Conservative government of William Davis in 1977. This was not the first time that such a department existed. In the 1920s the Conservative administration of Howard Ferguson had established the Department of Northern Development, but it did little more than administer a small northern development fund set up in 1912 to assist settlement and road construction in Northern Ontario. It was also short-lived, a casualty of the Depression and the cost-cutting measures of Liberal Premier Mitch Hepburn. Its successor was much more ambitious. Under the direction of its first minister, Leo Bernier, the self-proclaimed "King of the North," and his Liberal and

NDP successors, it has become the well-spring of provincial assistance for local community development.

Partnership between government, communities and the private sector is now a common phenomenon in Northern Ontario, with a plethora of government programs to channel development monies into the region. The presence of government has become even more pronounced with the transfer of several governmental offices from southern Ontario and Ottawa to the major cities and towns in the area in the past decade and a half. Sudbury was made the seat of the Ministry of Northern Development and Mines, as well as of the Ontario Geological Survey, and of one of the federal taxation data centres. Similarly, the federal government has relocated the Ontario regional office of the Department of Veterans Affairs to Kirkland Lake, and is proposing to establish an area goods and services tax office there as well.

To some degree these efforts to bolster the fortunes of Northern Ontario's mining communites reflect a belated recognition by governmental authorities of a fundamental inequity in Ontario's development process. Wealth generated in one part of the province, from the mines of Northern Ontario, has been drawn off and used to develop another-the commercial, industrial and financial heartland in southern Ontario. Moreover, because Ontario's peculiar assessment regulations historically restricted municipalities' right to tax mining companies, the lion's share of tax revenues went to the federal and provincial governments and never returned to the North. In one sense, then, the increasing governmental presence in Northern Ontario may be seen as a legitimate attempt to reverse a long-standing and unjust draining of wealth from the region.

THE INORD CONFERENCE

Given the central role that mining and mining communities have played in the life of Northern Ontario, the topic was a natural choice as a theme for the second annual conference of the Institute of Northern Ontario Research and Development (INORD) held at Laurentian University on 19 and 20 October 1990. As with the first INORD conference, which dealt with the contentious Temagami land-use issue, the aim was to bring together academic researchers from different disciplines and from all across Canada to discuss the key issues affecting mining and mining communities in Northern Ontario, past and present. The topics that were considered included the history of mining and mining communities, single-industry dependency, the environment, health and safety, and the role of government. Adding a sense of realism – and urgency – to the debate, yet still concluding the conference on a positive note were presentations by the municipal heads of the three communities most severely affected by recent mine closures: Elliot Lake, Kirkland Lake and Temagami. Intellectually the meeting of academic theory with practical experience proved to be both stimulating and rewarding.

THE PAPERS

Collectively and individually the papers published here address subjects common to the Northern Ontario mining community experience. First and foremost, of course, is the issue of dependency upon single-resource industries, and in some cases, upon single companies. How and why that dependency occurred and evolved is critical to an understanding of Ontario's northland. Ultimately, it underlies most of the problems of economic and social instability which have too frequently recurred in Northern Ontario communities, and which are also examined in the collection. It also points to the need for economic diversification and raises the question of how to achieve it for individual cities and towns and for Northern Ontario as a whole. Whether there is a formula for successful diversification that can be applied to communities across the North is a debatable question.

Another theme running through many of the papers presented here is the ever-increasing role of government in Northern Ontario mining communities, especially in the post-1945 era. While that was only one among several topics emphasized at the INORD conference itself, in the revised papers submitted for publication it emerged unsolicited as a provocative motif. Whether Northern Ontario genuinely differs from other parts of the province in its reliance on governmental support needs to be more fully examined. If it does, what does this say about the present and future status of Northern Ontario mining communities? Do they really deserve such assistance? If the original economic basis of a community disappears, should it be artificially maintained, or simply allowed to die a natural death? Is there a better alternative, such as fly-in mining, for natural resource development? Who should decide?

The papers in this collection make no pretence of offering definitive answers to these questions. Their objective, like that of the INORD conference itself, is much more modest: to provide the reader with a better understanding of the issues – past, present and future – that are critical to the Northern Ontario mining community experience.

ACKNOWLEDGEMENTS

In editing these manuscripts we have attempted to change as little as possible, aiming only to impose a reasonable consistency of style and usage on the various articles. With respect to documentation of sources, we have asked the authors to use the notational system most appropriate to their discipline.

We are indebted to the Social Sciences and Humanities Research Council and the Ontario Heritage Foundation (OHF) for the financial subventions they have provided for the publication of this volume. Particular thanks for their advice and editorial assistance are due to Paul Bator and Paul Litt of the

OHF, and Kirk Howard and Nadine Stoikoff of Dundurn Press. We must also acknowledge the contributions made by Lucille Brisebois, Emma Gagnon and Jane Pitblado both to the organization of the October 1990 conference, and to the preparation of this manuscript. Mr. Leo LaRiviere of Laurentian University's Department of Geography prepared some of the maps. Finally, we want to thank Laurentian University, the Institute of Northern Ontario Research and Development, and its director, Anne-Marie Mawhiney, for providing us with the resources and encouragement to complete this work.

Contributors

Matt Bray is associate professor of history, out-going director of the Institute of Northern Ontario Research and Development at Laurentian University, and co-editor of *Temagami: A Debate on Wilderness* (Toronto: Dundurn Press, 1990).

George Farkouh is mayor of Elliot Lake.

Eileen Goltz is an associate librarian at Laurentian University with a doctorate in history who has written several articles on Northern Ontario communities such as Copper Cliff and Espanola.

Peter V. Krats teaches history at the University of Western Ontario and is author of several articles relating to the Finns in Northern Ontario.

Joe Mavrinac is mayor of Kirkland Lake.

Ron Prefasi is reeve of Temagami.

Robert Robson teaches history at the University of Manitoba and is author of *Canadian Single Industry Communities: A Literature Review and Annotated Bibliography* (Sackville, N.B., 1986).

Oiva Saarinen is associate professor of geography at Laurentian University and author of numerous articles on communities in Northern Ontario.

Mark Shrimpton is a senior research fellow with the Institute of Social and Economic Research and adjunct professor in geography at Memorial University.

Keith Storey is director of the Offshore Oil Project of the Institute of Social and Economic Research and professor of geography at Memorial University.

Ashley Thomson is an associate librarian at Laurentian University and co-editor of *Temagami: A Debate on Wilderness* (Toronto: Dundurn Press, 1990).

Donald Trotter is a professor of mining engineering at Laurentian University whose numerous publications have been mainly in the field of mining health and safety.

Brian Walker is an assistant professor of sociology at the University of Toronto.

CM. *Wallace* is associate professor of history at Laurentian University and editor of "City Government in Northern Ontario," a special issue of the *Laurentian University Review* 17, no.2 (February 1985).

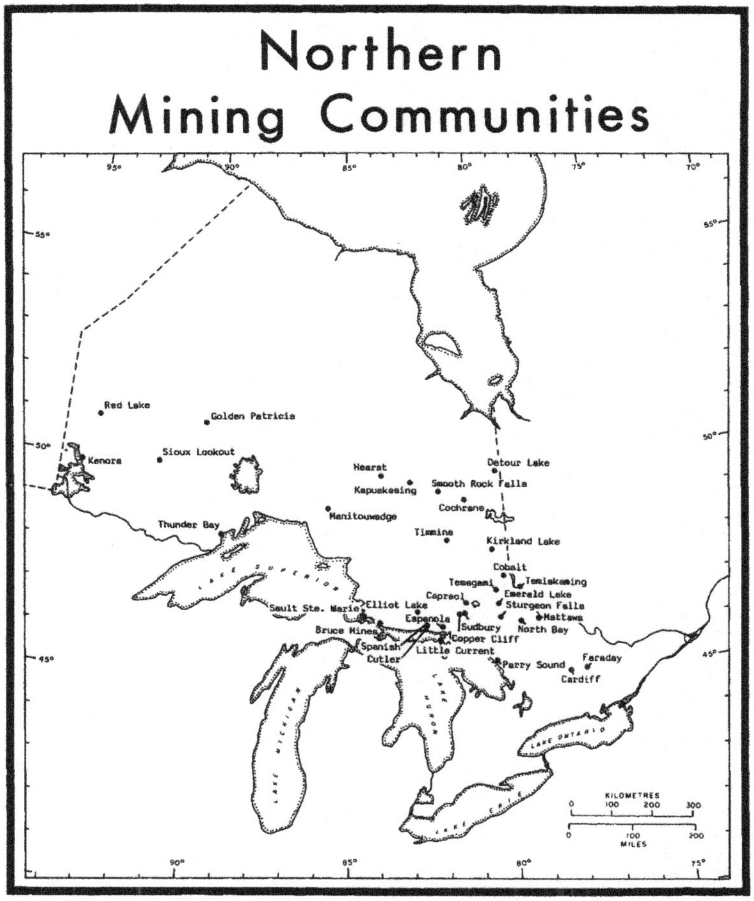

Courtesy of Léo Larivière

Part One

Setting the Stage

Courtesy of Lionne Photography

A modern community: Sudbury in the 1990s.

Introduction

Introducing this collection is a thought-provoking piece by Laurentian University's CM. Wallace. In "Communities in the Northern Ontario Frontier," Wallace, an associate professor of history, challenges conventional interpretations of Northern Ontario as a region of fragile one-industry towns and cities. While acknowledging that many northern communities were at first resource-based, he argues that most are now much more complex than their origins would indicate. He suggests that if they were compared to communities in other parts of Canada, north and south, some surprising similarities might emerge. At the same time, each community merits a detailed study of its own individuality and uniqueness. Thus, he demands a more sophisticated look at Northern Ontario communities, one which emphasizes not only the impact of external influences, but also the effects of their idiosyncratic internal forces. Wallace raises important general questions that can be asked of each of the specialized studies to follow, enriching our appreciation of their particular subjects and highlighting their thematic unity. In many ways he poses a provocative challenge not just to contributors to this volume, but to all students of urban history.

Chapter 1

Communities in the Northern Ontario Frontier

C.M. Wallace

The Laurentian Shield in Northern Ontario has spawned both a national and an international image of Canada.[1] Its essence is the blue wilderness lake, surrounded by shivering pine and birch, the tranquillity interrupted only by the haunting cry of the loon or the whispering paddle of the solitary canoeist. That was a reality captured by Tom Thomson and the Group of Seven, the only Canadian artists widely recognized in the country. The Temagami wilderness today is at the centre of a rearguard action by those who would preserve it as a relic of that world. Until the 1880s Northern Ontario was that wilderness, and while shadows of it remain, the towns and cities in the Shield have altered its texture.

Academics have found it convenient to describe the communities in Northern Ontario in simple categories such as "single-resource," "resource-based," "one-industry," "single-sector" and "company towns." Superficially appropriate for many communities, this categorization has constricted attempts to understand the urban North. Such nomenclature inhibits understanding of complex towns and cities now entering their second century. Like the image of the Group of Seven, that of the one-dimensional town of the North is a stereotype with severe limitations. This paper makes some observations on those communities and the study of them.

THE ORIGINS OF NORTHERN ONTARIO TOWNS

Until the arrival of the railway in the 1880s, Northern Ontario was remote, even mysterious. Except for the legendary intrusions of the fur traders, it was touched by civilization only on the fringes of Lakes Huron and Superior and Hudson Bay. The construction of several railroads through the Shield between 1880 and 1920 effected a transformation. The railways not only traversed the area, they created their own communities within it and made others possible.[2]

Table 1.1 Incorporation History: Northern Ontario 1880-1920

Community	Region		Incorporated as	
		Village	Town	City
North Bay	Nipissing	1882	1890	1925
Kenora	Kenora	1882	1892	
Port Arthur	Thunder Bay		1884	1907
Sault Ste. Marie	Algoma		1887	1912
Parry Sound	Parry Sound		1887	
Sundridge	Parry Sound	1889		

Community	Region		Incorporated as	
Little Current	Manitoulin		1890	
Gore Bay	Manitoulin		1890	
Burk's Falls	Parry Sound	1890		
Fort William	Thunder Bay		1892	1907
Mattawa	Nipissing	1892		
Sudbury	Sudbury		1892	1930
Thessalon	Algoma		1892	
Sturgeon Falls	Nipissing		1895	
Rosseau	Parry Sound	1896		
Copper Cliff	Sudbury		1901	
Bruce Mines	Algoma		1903	
Cache Bay	Nipissing		1903	
Fort Frances	Rainy River		1903	
New Liskeard	Temiskaming		1903	
South River	Parry Sound	1903		
Haileybury	Temiskaming		1904	
Massey	Sudbury		1904	
Powassan	Parry Sound		1904	
Rainy River	Rainy River		1904	
Steelton	Algoma		1904	
Webbwood	Sudbury		1906	
Blind River	Algoma		1906	
Bonfield	Nipissing	1906		
Cobalt	Temiskaming		1906	
Kearney	Parry Sound		1907	
Latchford	Temiskaming		1907	
Keewatin	Kenora		1908	
Englehart	Temiskaming		1908	
Chelmsford	Sudbury		1910	

Community	Region		Incorporated as	
Cochrane	Cochrane		1910	
Dryden	Kenora		1910	
Timmins	Cochrane		1912	1973
Sioux Look-out	Kenora		1912	
FroodMine	Sudbury	1913		
Trout Creek	Parry Sound		1913	
Charlton	Temiskaming		1914	
Iroquois Falls	Cochrane		1915	
Thornloe	Temiskaming		1916	
Capreol	Sudbury		1918	
Magnetawan	Parry Sound	1918		
Kapuskasing	Cochrane			1921
Hearst	Cochrane			1922
Hilton Beach	Algoma		1923	
Nesterville	Algoma			1923
Smooth Rock Falls	Cochrane			1929
Coniston	Sudbury			1934
Geraldton	Thunder Bay			1937
Larder Lake	Temiskaming			1938
Levack	Sudbury			1938
Lively	Sudbury			1953
Espanola	Sudbury			1958
Iron Bridge	Algoma		1960	
Elliot Lake	Algoma		1966	
Kirkland Lake	Temiskaming		1972	

Source: Elizabeth and Gerald Bloomfield with Peter McCaskell, Urban Growth and Local Services: The Development of Ontario Municipalities to 1981, University of Guelph, Department of Geography, Occasional Papers in Geography, no.3 (1983): 20-30 (with appreciation to O. Saarinen).

The Canadian Pacific was the first into the Shield, building its main line between 1882 and 1885. It added routes from Sudbury to Sault Ste. Marie in 1888 and from Sudbury to Toronto in 1908. By that time William Mackenzie and Donald Mann had constructed their Canadian Northern Railway from Port Arthur to Winnipeg, and extended it eastward across Northern Ontario by 1914. In the meantime the National Transcontinental had been pushed through the Shield. Several regional lines were also built, notably the Algoma Central north from Sault Ste. Marie to Hearst, and the Temiskaming and Northern Ontario Railway (T&NO) which eventually connected North Bay to James Bay and was renamed the Ontario Northland Railway. The incorporation history of northern towns indicates that approximately 85 percent of all communities in Northern Ontario were founded and/or incorporated between 1880 and 1920. Practically all of them were on a railway line. A few existed before 1880, such as Mattawa, Sault Ste. Marie, Port Arthur and Kenora, though none had achieved village status and most were in decline. (See Table 1.1 and Map 1.1.) This was the era, therefore, in which most of the significant urban places in Northern Ontario were founded.

To stress the importance of the railway in the development and character of Northern Ontario is to state the obvious, for its role has been accorded almost mythological dimensions. As recently as 31

March 1989, a *Toronto Star* headline declared: "For Northern Ontario folk the train is a way of life." In reality very few residents have actually travelled on the trains in recent years, though they want them retained as mechanical pets. The railway itself did not spur unlimited development in an area with sparse agricultural land and timber stands that were soon to be exhausted.

The opening of the North coincided with several scientific and technological innovations that go under the heading of the Second Industrial Revolution.[3] A combination of applied innovations in physics, chemistry and engineering permitted the rapid exploitation of northern resources and the growth of communities. The industrial application of electricity and chemistry to mining and pulp and paper engineering unleashed the economic forces which drew the map of urban Northern Ontario. Even in farming, government scientific experimental stations produced much of the know-how that expedited agriculture in areas such as the clay belt.

The role of governments, federal and provincial, was another important determinant. The Canadian Pacific Railway (CPR), National Transcontinental and the Temiskaming and Northern Ontario Railway (T&NO) were all creations of government in one way or another and were built with government assistance. The T&NO, for example, was constructed from North Bay to Cochrane between 1902 and 1908 by the Province of Ontario as part of a conscious policy designed to exploit the resources of the area. Intended especially to provide access to minerals and forests, it was also to serve as a colonizing road to establish settlements.[4] The numerous mining and pulp and paper communities that were established between 1903 and 1916 fully justified the government expenditure on railroads, especially after the gold and silver from places like Cobalt, Timmins and Kirkland Lake generated both private and public wealth. Ontario Hydro, another government agency, quickly moved north and was of critical importance in the exploitation of the natural resources. From the beginning the Ontario government maintained more than a passing interest in its North and its communities, especially in the areas of local governments and the services essential to resource development that they provided.[5]

Map 1.1 Railways in Northern Ontario

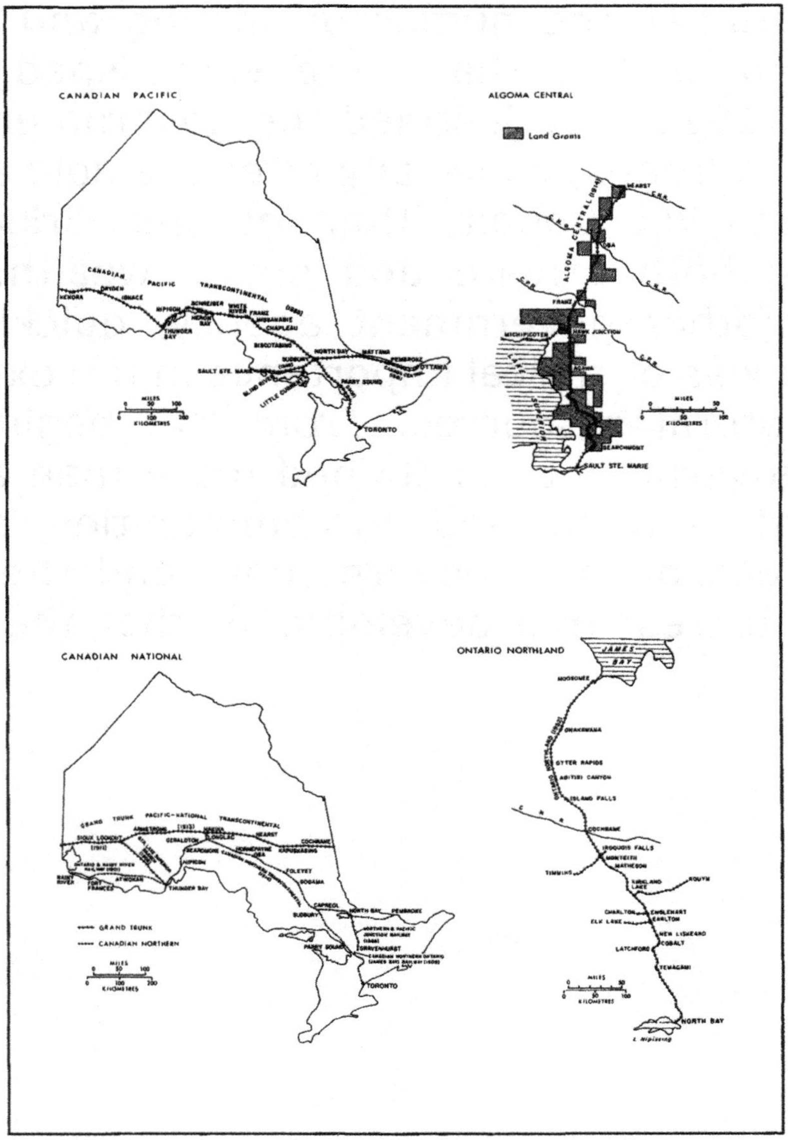

Source: Department of Geography, Laurentian University, and O.W. Saarinen.

THE NORTHERN TOWNS AND THE ACADEMICS

That the towns in Northern Ontario would be considered "frontier" was probably inevitable. The tendency of academics has been to lump all towns together into categories, usually in time/form phases such as "unplanned camp sites," "unregulated growth," "planned townsites" and so on. Most of the towns in Northern Ontario undoubtedly began as resource-based, one-industry communities constructed in the wilderness to serve the needs of a company or companies. They were "colonial towns," forming part of what G.A. Stelter has called the Canadian "urban frontier."[6]

The concept of the one-industry town as the "spearhead" of that frontier is favoured by historians. Stelter and Alan Artibise presented this thesis most succinctly in their article, "Canadian Resource Towns in Historical Perspective," for *Plan Canada* in 1978. In that exploratory study the towns were seen to pass through several distinct stages based on planning and function. A feature of their approach was that the towns were studied as similar historical objects. They identified four common characteristics: the towns were adjuncts to an industrial empire; they lacked control over their development; they had a simplified occupational structure; and they shared a common physical appearance. Examples of changes in form and function over time were explored, with the emphasis on the role and nature of planning.[7]

Historians, however, have been bit players on an academic stage dominated by geographers. They have written the majority of the studies and have found it convenient to categorize the towns as "single enterprise," "dependent," "single-sector" or as "hinterlands" to the metropolis. Ira Robinson's *New Industrial Towns on Canada's Resource Frontier* (1962) set the pattern for subsequent studies, both academic and administrative.[8] At the centre of much of the writings by geographers has been planning, or the lack of it, and this approach has come to dominate their writings, especially studies such as N.E.P. Pressman's *Planning New Communities in Canada* (1974) and Larry McCann's "Changing Internal Structure in Canadian Resource Towns," *Plan Canada* (1978).

McCann's interests, however, extended beyond planning and the communities themselves. In "Canadian Resource Towns: A Heartland-Hinterland Perspective" (1980), he has written the most sophisticated study yet to appear.[9] Placing the resource towns solidly within the metropolitan empire, he argued that they all followed similar patterns of development – economically, physically and socially. Their function as "colonial outposts of an industrial heartland" predetermined their evolutionary pattern and form. Oiva W. Saarinen picked up several of these themes in "Single-Sector Communities in Northern Ontario: The Creation and Planning of Dependent Towns" (1986).

He provided concrete examples of planning and development throughout the North and offered this analysis:

> This close dependency upon the natural environment has given the northern communities a number of common structural features which include small populations, slow growth rates, isolation, limited hinterlands, a narrow economic base, and poorly developed physical and sociocultural infrastructures. As well, the functioning of the urban network as a whole has been seriously influenced by powerful exogenous forces, instability and uncertainty, and weak spatial interaction among the constituent communities. These elements have all combined to create poor future growth potential.[10]

Much of this is echoed in Robert Robson's "Canadian Resource Towns: An Historical Overview, 1880-1970,"[11] though Robson's work is less theoretical and more solidly historical than other studies.

Categorization is necessary and contributes to our understanding of the Canadian system of cities, but it does have its limitations. All towns of Northern Ontario, whether resource-based or not, have become caught up in a problem-solution paradigm. The essence of the studies is the problem with planning, the problem with development and servicing, and the problem with external relationships and government policy. The unstated assumption is that the towns themselves are problems to be solved, rather than complex communities to be understood. A similar conclusion was reached by Robson in his *Canadian Single Industry Communities: A Literature Review and Annotated Bibliography* (1986), a recent compilation with a thorough summary essay.[12]

An obvious shortcoming of most studies is that they concentrated on the initial or "threshold" stage in the life of the communities. This is especially the case with studies that centre on early development and planning, a bias that develops naturally from using external sources. The result is stylized works that present the communities as lifeless structures. Every urban centre in Northern Ontario is stereotyped as a single-enterprise frontier town, doomed to extinction when the resources run out. It is worth noting that few studies actually follow these communities to their extinction,[13] and there are virtually no studies on the nature and functioning of any institutions or structures over time in these communities.

Building on the existing literature, it is appropriate to rethink the history of communities in this Ontario frontier. An essential beginning is to abandon the concept of frontier towns, except for the threshold phase, and to treat them as "normal" towns and cities. Many have been urban places for upwards of a century. Hand in hand with this, academics should end their preoccupation with problems, unless it would be to compare the problems with other non-northern communities and turn to other issues. The list of "limitations" set out by Saarinen above is much more revealing and less overwhelming when placed in juxtaposition with the disadvantages of most non-central-place towns.

Geographers like Saarinen – all researchers for that matter – have been supplementing our scanty knowledge. Historians, in particular, need to know more about these towns as individual communities. Very few Northern Ontario towns and cities have been studied extensively, and most works that exist are poor antiquarianism. When an issue of the *Laurentian University Review* on "City Government in Northern Ontario" was put together in 1985, the authors of every article had to reconstruct even the elementary political history of each community from primary sources because so little was known about its form of government, its people, its politics, or, for that matter, anything else. When seeking information on social structures, business organizations, services or social activities, researchers entered a wasteland. This explains why so much emphasis has been on planning and the heartland-hinterland relationships. This information is readily available, but it is all external to the actual towns.

There have been exceptions where the communities were studied internally. Stelter, for example, wrote two articles on early Sudbury,[14] and Doug Baldwin has done several on Cobalt.[15] Eileen Goltz has written on "Espanola: The History of a Pulp and Paper Town,"[16] and on Copper Cliff, "Genesis and

Growth of a Company Town: 1886-1920."[17] Michael Kelly examined "Sudbury as a Regional Metropolis, 1900 to 1910."[18] What is most striking about these studies when they are examined as a group is not so much the similarities they reveal as the diversity in development and differences among northern and company towns. Even a cursory examination of the cities of North Bay, Timmins, Sudbury, Sault Ste. Marie, Port Arthur and Fort William reveals that each has had a distinct pattern of settlement and development, and each has had very different functions. Only Timmins and Sudbury are considered resource-based in the limited sense, and Sudbury grew into a very complex city. Sturgeon Falls, Espanola and Kapuskasing, all smaller forestry or mill towns, are distinguished from each other by their unique historical patterns, including differences in their origins, growth and development, economies, roles, ethnic compositions, and external relations. All of this is lost in the rush to generalize from externals such as planning and corporate control.

At a glance they appear to personify Rex Lucas's *Minetown, Milltown, Railtown* communities with "few past memories ... no lingering myths of days gone by ... [whose] future depends upon impersonal forces outside their community."[19] Only by ignoring or not knowing the unique imprint of each community is it possible to come to such a conclusion. Each has its own relationship with outside ownership, its own booms and busts, ethnic encounters, political leaders, and religious diversity, and it is these historical accidents that make it distinctive and provide its mythology. A preoccupation with detail and uniqueness will no doubt lead to the charge of antiquarianism, but as things stand now, the generalizations based on externals err on the side of superficiality.[20]

SOME NEW APPROACHES

It is time to reconstruct the history of the Northern Ontario communities. Instead of lumping them all together in geographic/time/form categories and falling into Lucas's trap of starting out knowing the answers, academics must ask different questions and consider different perspectives. From the basic information on the cities and towns that has been collected, it is more appropriate to compare them, not just to others in the frontier, a proscription from the start, but to communities of their era and size from across Canada and abroad and on a wide spectrum that takes into account more than just planning.[21] The role of elites and boosters, for example, is a fertile subject for comparison, as is the nature of political leadership. Despite Lucas's opinion that interest in local government was "casual,"[22] the evidence for Cobalt, North Bay, Timmins, Sudbury and Sault Ste. Marie, to name just a few examples, suggests a vigorous participation.

Each town produced several generations of local political and business leaders, some of them quite dynamic, who were extensions of their communities.[23] Frank Cochrane of Sudbury, for example, moved easily from being a local entrepreneur and politician to the corridors of both big business and national politics. W. Marr Brodie went directly to the Ontario Municipal Board in the mid-1930s on the basis of his effective crisis management as mayor of Sudbury. More recently, during the 1960s and 1970s, Timmins, North Bay, Sault Ste. Marie and Sudbury all had colourful, populist mayors who maintained their positions over time with a variety of skills and well-oiled machines. Leo Del Villano of Timmins, Merel Dickerson of North Bay, Joe Fabbro of Sudbury and, to a lesser extent, John Rhodes of Sault Ste. Marie had much in common. All were typical of political leadership in many industrial or commercial towns of the time. Windsor, Trenton, Barrie and Oshawa had comparable types of politicians. The members of the city councils, the school boards, the chambers of commerce, the Rotary and Lions clubs are interchangeable among most of these cities in terms of occupation, ethnicity, education and other defining characteristics.

The argument here is that the type of questions asked in the past predicted the types of answers received, and that inappropriate questions may have been asked in the first place. It is not beyond the realm of possibility, to press a point, that there might be a place for some individuals, other than the planner Thomas Adams, in the history of Northern Ontario.

A preliminary historical survey of business establishments in Timmins, North Bay, Sudbury and Sault

Ste. Marie indicates that their range was consistent with similar communities in Canada.[24] The corporations in resource communities behave in a similar manner as those in other sectors, and their manipulations of politicians at all levels is of the same quality. The ethnic composition, religious activity and education of northern towns do not appear unusual when placed beside a cross section of Canadian communities. These conclusions, based on preliminary findings, will no doubt be revised, yet they do suggest that the towns in the Northern Ontario frontier are multidimensional.

The mentality of people in northern communities is another theme in need of revision. Terms such as "northern alienation," "cabin fever" and "manly north" have been fodder for journalists and academics alike. They "know" they are appropriate because they have "seen" them. The uniqueness of northern existence is taken for granted and rhetorical politicians find cliches useful in their pursuit of handouts. The alienation of northern communities is a favoured concept, but is it unique or different from that experienced in all non-metropolitan communities, or does it actually exist? Sociological literature indicates that alienation is universal. Evidence that it may exist in Northern Ontario does not suggest that the area is unique. It may demonstrate the reverse: that Sault Ste. Marie and Kingston, for example, share a relationship with the external world. Though cabin fever is usually related to geographic isolation, it is a vague concept.

It is also experienced, in the view of some psychologists, in many locations, including large cities. Other psychologists argue it is simply myth, with no application.[25] Finally, the concept of the manly north with the stereotypical hunter or fisherman sporting a baseball cap and driving a half-ton truck ignores the reality that most northerners neither fish nor hunt, or that it is urban dwellers from the south living out fanciful dreams who fill most of the boats on the lakes and keep hunting operators in business. Laurentian University's Elliot Lake Tracking Study has already produced significant results which should revise the thinking. In "Gender-Typed Activities in Mining Community Families: A Study of Elliot Lake in 1990," Derek Wilkinson used the evidence from 1, 182 questionnaires to show that, among other things, men in that community spend, on average, six hours per week on housework, which is significantly above the Canadian average for males.[26] Perhaps the "manly" is simply another cliche perpetuated by fuzzy thinking.

Through detailed studies sensitive to nuance, much of the uniqueness of the Northern Ontario communities may turn out to have been a myth. The image of Northern Ontario as an area filled with fragile one-industry towns will persist, as will the wilderness images of the Group of Seven. As those towns enter their second century, however, they deserve more than a cursory invocation of stereotypes. Some are experiencing their fourth and fifth generation of residents, buildings, services, businesses, bosses and planners.

It is time for academics to take a second look, to dig below the surface. It was below the surface, after all, that the real wealth of Northern Ontario was found.

Notes

[1] An earlier version of this paper was presented at the conference on "The Nineteenth Century Canadian City: Internal Change and External Links," Centre of Canadian Studies, University of Edinburgh, 5 May 1989.

[2] The role of the railway in the development of Northern Ontario has produced an extensive literature, ranging from interpretative early articles such as Harold Innis, "Transportation as a Factor in Canadian History," *Proceedings of the Canadian Political Science Association,* 1931, 166-84, to detailed works such as O.S. Nock, *The Algoma Central Railway* (1976). See *Bibliography: History of North-Eastern Ontario* (1985), 1021-72. Innis wrote *A History of the Canadian Pacific Railway* (1933) to which Pierre Berton with *The National Dream* (1971) and *The Last Spike* (1974) added colour, especially for Northern Ontario. For the many lines that became the Canadian National, see G.R. Stevens, *Canadian National Railways, 2* vols. (1960 and 1962). Albert Tucker, *Steam into Wilderness: Ontario Northland Railway, 19021962* (1978), is useful, though like all the studies is concerned with the railroads rather than their impact on communi-

ties. See the map from Oiva W. Saarinen, "Single Sector Communities in Northern Ontario: The Creation and Planning of Dependent Towns," in Gilbert A. Stelter and Alan F.J. Artibise, eds., *Power and Place: Canadian Urban Development in the North American Context* (Vancouver: University of British Columbia, 1986), 223.

[3] A.R.M. Lower, among others, used this concept. See his *Settlement and the Forest Frontier in Eastern Canada* (Toronto: Macmillan, 1936), 114. It is well established in scientific and technological literature. See, for example, James Kip Finch, *The Story of Engineering* (New York: Doubleday, 1960). Diane Newell has applied it to this area for mining in *Technology on the Frontier: Mining in Old Ontario* (Vancouver: University of British Columbia, 1986).

[4] See Tucker, *Steam into Wilderness,* and the issue of the *Laurentian University Review* (February 1981) on the "Ontario Northland Railway," edited by G.S. Mount and A.D. Gilbert.

[5] See Oiva Saarinen, "Municipal Government in Northern Ontario: An Overview," in "City Government in Northern Ontario," *Laurentian University Review* (February 1985), edited by CM. Wallace.

[6] "The Urban Frontier in Canadian History," in A.R. McCormack and I. Macpherson, eds., *Cities in the West: Papers of the Western Canadian Urban*

History Conference (Ottawa: National Museum of Man, 1975).

[7] A revised version published in their *Shaping the Urban Landscape: Aspects of the Canadian City-Building Process* (Ottawa: Carleton, 1982), 413-34, was used here.

[8] Robinson emphasized the "resource" aspect, but he relied heavily on *Single-Enterprise Communities in Canada* (Ottawa, 1953), prepared by the Institute of Local Government, Queen's University.

[9] In Richard E. Preston and Lome H. Russwurm, eds., *Essays on Canadian Urban Process and Form* (Department of Geography, University of Waterloo, Kitchener, Publication Series no.15, 1980).

[10] In Stelter and Artibise, eds., *Power and Place,* 228.

[11] An unpublished paper delivered at a Canadian Studies Conference, University of Lund, Sweden, 11-14 August 1987. Robert Robson's, "The Politics of Resource Town Development: Ontario Resource Communities, 1883-1970" (Ph.D. thesis, University of Guelph, 1986), was not available at the time of writing.

[12] Robert Robson, *A Literature Review and Annotated Bibliography* (published by the Rural and Small Town Research Studies Programme, Department of Geography, Mount Allison Univer-

sity, Sackville, New Brunswick, December 1986).

[13] An almost unique exception is Robert Robson, "Ontario Hydro Colonies: A Study of Frontier Settlements," *Laurentian University Review* 17, no.2 (February 1985): 113-39. Robson examined Cameron Falls, Ear Falls and Abitibi Canyon.

[14] "The Origin of a Company Town: Sudbury in the Nineteenth Century," *Laurentian University Review* (February 1971), and "Community Development in Toronto's Commercial Empire: The Industrial Towns of the Nickel Belt, 1883-1931," *Laurentian University Review,* June 1974.

[15] See "The Development of an Unplanned Community: Cobalt 1903-1913," *Plan Canada,* March 1978, and "A Study in Social Control: The Life of the Silver Miner in Northern Ontario," *Labour/Le Travailleur 2* (1977). Robson, *Canadian Single Enterprise Communities,* has a complete list.

[16] Eileen Goltz, "Espanola," *Laurentian University Review,* June 1974.

[17] Eileen Goltz, "Genesis and Growth of a Company Town" (M.A. thesis, Department of History, Laurentian University, Sudbury, Ontario, 1983).

[18] Michael Kelly, "By Divine Right: Sudbury as a Regional Metropolis, 1900–1910" (M.A. thesis, University of Toronto, 1984).

[19] Rex Lucas, *Minetown, Milltown, Railtown: Life in Canadian Communities of Single Industry* (Toronto: University of Toronto, 1971), 20.

[20] After this paper was delivered at Edinburgh, there was a comment that it was going in the opposite direction of urban studies internationally. Urbanologists, it was argued, were looking for global generalizations, not more detailed studies such as was proposed in this paper. The pursuit of patterns and universals is not being questioned here; it is the base from which they start. Meta-urbanists, like meta-historians, must build on solid foundations. This paper suggests that in the case of communities in Northern Ontario that foundation needs more substance, and the one that has been erected may be flawed. The distinction drawn here is that noted by David Cannadine in "Urban History in the United Kingdom: the Dyos phenomenon' and after," in David Cannadine and David Reeder, eds., *Exploring the Urban Past: Essays in Urban History by H.J. Dyos* (Cambridge: Cambridge University Press, 1982), 216.

[21] This proposal, in particular, interested the commentator on this Edinburgh paper.

[22] Lucas, *Minetown, Milltown, Railtown,* 20.

[23] See, for example, John Lawrence Kesik, "A Linguistic and Religious Study of Sudbury Town

Council, 1893-1930" (Honours essay, Department of History, Laurentian University, 1985).

[24] See Maury O'Neill and Peter Andrews, *Sudbury Statistical Summary* (Laurentian University, Community Studies Project, 1986). Sandra Walton is preparing a companion summary for Timmins, Sault Ste. Marie and North Bay. See also Judith Topham, "Sudbury: Growth and Development, 1920-1940" (Honours essay, Department of History, Laurentian University, 1981), and Richard Philbin, "Growth and Development in North Bay and Area" (Honours essay, Department of History, Laurentian University, 1987). Holes are being filled by studies such as Duncan McDowall, *Steel at the Sault: Francis H. Clergue, Sir James Dunn, and the Algoma Steel Corporation, 1901-1956* (Toronto: University of Toronto, 1984).

[25] Even for the Yukon "those cabin fever yams are nothing more than myths," if psychologist Pat Kehoe is correct. See *Toronto Star,* 26 December 1988.

[26] Derek Wilkinson, "Gender-Typed Activities in Mining Community Families: A Study of Elliot Lake in 1990" (Paper presented to the Canadian Association of Sociology and Anthropology, Kingston, 4 June 1991).

Part Two

Early Years

Copper Cliff, looking towards Sudbury, 1903 and 1991.

Introduction

This section examines the formative years of Northern Ontario mining communities in the late nineteenth and early twentieth centuries, with particular emphasis on the Sudbury region. Its three articles address several critical questions about the nickel-copper mining industry and the towns to which it gave rise. What was a "company" town, and how, if at all, did it differ from other communities? Did its character change over the years or was it fixed in stone at the time of origin? What factors shaped the relationship between companies and their communities? What impact, if any, did the type of industry and/or the nature of the company have on that relationship?

In "A Company and a Community: The Canadian Copper Company and Sudbury, 1886-1902," Matt Bray, one of the co-editors of this volume and an associate professor of history at Laurentian University, discusses the Canadian Copper Company's (CCC) changing views of Sudbury between 1886 and 1902. It should be noted that Bray's paper was not delivered at the Institute of Northern Ontario Research and Development (INORD) conference, having been previously presented at the University of Edinburgh in May 1989, and that his reluctance to include it has been overcome by the insistence of others, including his co-editor, for two reasons. The first is that it presents new material drawn from the Inco archives, recently opened to academic researchers. The second, and more important, is that Bray's focus on the company perspective of the community rather than the converse is unusual and particularly relevant to the theme of this volume.

In "All That Glitters: Speculation and Development in the 'Minor' Minerals of the Sudbury Area," Peter Krats also contributes a fresh perspective on Sudbury's early history. Krats reminds us that despite the region's reputation as the "Nickel Capital of the World," the area was rich in many other minerals. Even though the search for such minerals was largely unsuccessful, Krats argues, "a maturing regional infrastructure owed much to the gold, lead and even the coal seeker."

In its own way, Eileen Goltz's "The Image and the Reality of Life in a Northern Ontario Company-Owned Town" more than meets the challenge set out by Wallace in the introductory paper to this collection. It is Goltz's view that outside the town of Copper Cliff, the Canadian Copper Company and later the International Nickel Company had relatively benevolent images that were not always reflected in the realities of the community itself. Goltz, an associate librarian at Laurentian University, has also been able to secure access to Inco's archives, and her contribution to this volume draws on previously unavailable sources.

Chapter 2

A Company and a Community: The Canadian Copper Company and Sudbury, 1886-1902

Matt Bray

In the last two decades, the study of single-industry towns in Canada has become a growth industry. Thanks to the work of historians and geographers such as Stelter, Artibise, McCann and Robson, to name only a few, we now have a much more sophisticated understanding of this particular form of urban place. Case studies of individual communities from Larder Lake to Nanaimo, and Espanola to Lethbridge have provided the basis for general analyses of the factors shaping the appearance of single-industry towns, their defining characteristics, the stages in their development and their perceptions of the particular industry on which they were dependent.[1]

The company-town relationship has generally been approached from the perspective of the community, as in the studies of the Canadian Copper Company (CCC), which became the International Nickel Company of Canada, Limited (Inco), and the Sudbury area by individuals such as Gil Stelter, Noel Beach and Eileen Goltz.[2] The company's vantage point has been more difficult to determine. To some extent this is because the subject falls more logically into the realm of business history, a branch of the discipline that in Canada is still somewhat in its infancy. It is also partly related to the problem of sources. Businesses have generally not been overly concerned about preserving and/or organizing their historical records, and they have been notoriously wary of giving historians access to whatever archives they do have.

A case in point is the International Nickel Company of Canada. For many years its North American archival holdings were closed, primarily because they were scattered at various offices in Ontario, New York and New Jersey. Fortunately, the adoption of new records management procedures in the last decade and a half has led to the centralization of Inco's holdings at Sudbury, and as a result the company has selectively opened its archives to historical researchers, facilitating an examination of the company-town relationship from the company's perspective.

This study is limited to the period from 1886 to 1902, the years when Canadian Copper, Inco's forerunner in the Sudbury district, was an independent, Cleveland-based, nickel-copper mining company. During these formative years ownership was concentrated in the hands of four Ohio men, Sam Ritchie of Talmadge, Thomas Cornell of Akron, and from Cleveland, Henry B. Payne and Stevenson Burke; the lines of authority between Canadian Copper's head office in Cleveland and local headquarters (in Sudbury from 1886 to 1890, and thereafter in Copper Cliff) were uncomplicated; and the decision-making process was relatively clear-cut. Both the responsibility and the rationale for policies and practices that shaped the company's attitude towards Sudbury during this period can, therefore, be determined with a great degree of certainty.

This paper also focuses on the attitude of Canadian Copper towards Sudbury, as distinct from Copper Cliff, which was more genuinely a company town in the conventional sense of that phrase.[3] Sudbury was in a somewhat different category, for while it was indirectly dependent upon mining, it was not "owned" by the CCC and was therefore able to maintain a measure of separateness. As Gil Stelter has discussed, Sudbury's attitude towards the Canadian Copper Company during this era was one of ambivalence.[4] As the largest and, indeed, the only operating mining company in the Sudbury area for much of the 1890s, Canadian Copper's authority engendered both respect and animosity in the community. In the pages of the *Sudbury Journal,* for example, there was an unease about the company that occasionally erupted into outright hostility, as in 1899 when the paper levelled a bitter attack on the CCC's environmentally devastating heap-roast smelting practices.[5]

Canadian Copper's attitude towards Sudbury traced a distinctive but equally ambivalent pattern during these years. For the first five years of its existence the company was reasonably considerate of Sudbury's interests, although in the late 1880s it did begin to distance itself somewhat from the community. A definite break in relations occurred in 1891, and the years from 1891 to 1898 were characterized by indifference and distrust on the company's part. At the end of the decade the company made a concerted effort to put its relationship with the community on a more harmonious basis once again, an effort which persisted until ownership passed into the hands of the International Nickel Company, a U.S. Steel-controlled holding company created in 1902.

Shaping Canadian Copper's equivocal attitude towards Sudbury were several locational, managerial and corporate constraints. Some of these were generic in nature, arising out of the CCC's position as the dominant company in a single-industry community. Others were unique, products of the conflicting personalities and interests of its major corporate owners. Both are critical to an understanding of the relationship between the company and the town in these years.

EARLY GOODWILL

With respect to the initial period of goodwill between Canadian Copper and Sudbury, location was a vital factor. From 1886 to 1890 the Canadian

Copper Company was closely identified with Sudbury because its local offices were situated there and its management personnel lived in the village and actively participated in its social affairs. The association would have become permanent had the company's original plan for its smelting operations been followed. On strictly business grounds Sudbury was considered "much the best place" for the smelter because it was there that the branch railway lines from its two separated mines at Copper Cliff and Stobie converged. This plan was not implemented because of concern about the effect on the community of the sulphur fumes that would inevitably be associated with the smelting process.[6]

The hope that locating smelter operations outside of Sudbury would avoid the emissions problem would quickly prove naive but, nevertheless, in the fall of 1888 the smelter complex was constructed at Copper Cliff, then the more important of the CCC's mines and closer to other prospective ore bodies. This, followed by the transfer to Copper Cliff of the company offices early in 1890, significantly altered the relationship between the Canadian Copper Company and its communities. In distance the move was only a few miles, but the consequences were much more far-reaching. Increasingly, management personnel, like the smelter workforce itself, not only worked but also lived in Copper Cliff. As studies by Gil Stelter and Eileen Goltz have demonstrated, then began Copper Cliff's evolution as a classic company town of the era.[7]

INDIFFERENCE

A distinct weakening of the psychological link between Canadian Copper and Sudbury also then began. The extent to which the company came to identify with Copper Cliff was manifested in 1901 when the company's assistant general manager, A.P. Turner, rationalized the application for the incorporation of Copper Cliff on the grounds that the company, rather than "indirectly keeping up Sudbury," should be "taking steps to build up our own town," Copper Cliff.[8]

Clearly an important part of the explanation for the CCC's increasing indifference towards Sudbury in the 1890s lay in its absentee ownership. Neither Thomas Cornell nor Stevenson Burke, its presidents during the decade, visited Sudbury very frequently or made any real effort to get to know its people. To a considerable extent this was because for neither of these men, nor for the third major shareholder, Senator H.B. Payne, was Canadian Copper their primary business.[9] In addition, two of these principal owners died during these years, Cornell in 1892 and Payne in 1896, leaving the handling of their interests in the company to estates which had still less attachment.

By late 1887 responsibility for the day-to-day running of the Canadian Copper Company fell mainly to its secretary-treasurer, Henry Payne Mcintosh, the namesake, protege and executive assistant of then senator Payne. Mcintosh was based in Cleveland, and his duties, according to a later company official, E.A. Collins, "included those usually devolving upon the managing director, purchasing agent, comptroller, secretary-treasurer, consulting metallurgist, chief engineer, and, by no means least, head of the Court of Appeal to settle disputes and incipient wars among those in charge at Copper Cliff."[10] A tough, demanding taskmaster,[11] he put in place at Copper Cliff organizational and administrative systems that contributed significantly to the company's rapid and long-lasting success. Mcintosh considered it his job to keep absolute control over all phases of the CCC's business. First and foremost that meant he had to be kept fully informed at all times. Lengthy daily correspondence flowed from Copper Cliff to Cleveland minutely detailing every aspect of the local operations. Back from Cleveland came equally long and detailed instructions about policies and practices to be followed. Little of the decision-making process was left to the local authorities.

More than any other company official, H.P. Mcintosh set the tone of Canadian Copper's attitude towards Sudbury during the 1890s, and that attitude was guarded, to say the least. Perhaps the best illustration of the suspicion with which he viewed the community was his obsession with secrecy about all aspects of the company's operations. This concern for protecting the company from prying eyes was first manifested in 1888 when he refused a request from the American Science Association to visit the mines on the grounds that "such a visit will be more likely to damage us by divulging our business."[12] A year later, again on Mcintosh's instructions, access was prohibited to local Sudbury visitors as well.[13] With the construction at Copper Cliff in 1891 of the Bessemer plant, designed to further concentrate the nickel-copper matte by the elimination of its iron content, this concern for secrecy became a passion. As General Manager John D. Evans explained to Mcintosh at the beginning of 1892, "we give no permits [of admission] whatsoever to the Bessemer plant. Over every door there is a sign about three feet large with the words 'NO ADMITTANCE' and we have men to keep watch on strangers about the place as also even our own men."[14]

Keeping unwanted visitors out of the mines and smelter was relatively easy. More difficult was keeping information about the company's operations, production, sales and so forth out of the public domain. Even the Canadian census in the summer of 1891 was perceived as a danger, and every effort was made to follow Mcintosh's instructions to "stave off" the census taker.[15] A more serious long-term threat appeared on 5 March 1891, the date of the first edition of the weekly *Sudbury Journal* edited by James A. Orr. In June of that year, Francis Sperry, a chemist who had recently left Canadian Copper, asked an acquaintance in Toronto to inform the press there about the company's sale of matte to the U.S. Navy. Sperry explained that in this way Orr would be able to legitimately reprint the information in the *Journal.* The problem, he said, was that "Evans takes Mr Orr to task every time a local item appears in the *Journal* about the CCC's affairs. Even going so far as to drive up and ask him how he obtained the information."[16] Sperry was absolutely correct. Evans's diary has numerous references to his doing just that, going to Sudbury to "interview" Orr about what had appeared in the paper.[17]

Neither from the perspective of the Canadian Copper Company nor from that of the *Sudbury Journal* did the problem disappear over the course of the 1890s. Early in 1896, for example, Orr complained bitterly in the *Journal* about the refusal of the mining companies in the area to divulge any information whatsoever about their operations; determining their profitability was particularly difficult, he noted, because "they are all close [*sic*] corporations, and no one can find out what their balance sheets tell at the end of the year."[18] For Canadian Copper, on the other hand, too many reports of its activities appeared in the press. The company's assistant general manager, J.N. Glidden, told Mcintosh in February 1898 that he was still trying to discover who was supplying Orr with all of his information. "There is no doubt," he wrote, "that McCharles [Aeneas McCharles, a local independent mining entrepreneur and long-time critic of the company] furnishes the most of it but the late article on profits indicates that he is picking up information from sources which it would be well for us to locate."[19]

Mcintosh's obsession with secrecy about company affairs was attributable to the fact that Sudbury was not solely a Canadian Copper town, for while the CCC was the dominant nickel producer in the district, it was not the only one. The years from 1889 to 1900 witnessed a succession of rivals – H.H. Vivian and Company of Wales, the Canadian Pacific Railway-backed Dominion Mineral Company, the Drury Nickel Company of Chicago, the Consolidated Lake Superior Corporation of Sault Ste. Marie, and the British Mond Nickel Company.[20] Only Mond Nickel would eventually become a serious competitor to the CCC, but the others posed real enough threats to cause it considerable anxiety.[21] This apprehension did not even significantly diminish in 1895 when CCC's status as a major player on the international nickel scene was assured by the market-sharing agreement which, along with the Orford Copper Company, its New Jersey refining partner, it negotiated with the France-based New Caledonian nickel giant, Le Nickel.[22]

Canadian Copper's ambivalent attitude towards Sudbury was also inspired by the fact that from a strictly business point of view the town had very little to offer the company. In the case of management personnel, for example, Frederick Bernhard, a company clerk in the late 1890s, remembered that as long as the company was Cleveland-based, all appointments from the master mechanic upward were personally made by Mcintosh from head office and not by local Copper Cliff officials.[23] When combined with the company's rapid turnover of personnel, this meant that the association between individual company officials and the community was seldom very close. For example, the first general manager, Lewis Ashmun, lasted from 1886 to 1888 and was succeeded by E.D. Peters, who left in the spring of 1890. He in turn was replaced by L.S. Woodbury of Michigan, who departed after only three months, followed by John D. Evans, who served until June of 1893.[24] For the rest of the decade the office was nominally filled by James McArthur, the former smelter superintendent, but much of the administrative work fell to Assistant General Manager J.N. Glidden until 1898, and thereafter, A.P. Turner.

The turnover in general managers was paralleled in other staff positions. Filling the position of company chemist vacated by Francis Sperry early in 1891 proved to be a major difficulty, as illustrated by the following series of excerpts from the diary of John D. Evans:

February 11, 1891	Mr Stark arrived from Cleve-land to take Sperry's place.
February 13, 1891	Mr Stark left for Cleveland.
March 11, 1891	Mr Leonard arrived from Cleveland to take up chemist's duties.
March 13, 1891	Mr Leonard left after dinner.

April 25, 1891	Mr Browne, the new chemist, arrived in the morning.
September 30, 1892	Mr Parks, Mr Browne's successor, arrived.[25]

The same pattern was traced with regard to bookkeepers. The first, Tom Kirkwood, was succeeded by a Mr. Land, who was succeeded by a Mr. Paull, who was succeeded by a Mr. Shiels (who died of consumption barely a week after arriving), who was succeeded by Mr. Glidden. The lack of continuity in these various positions was disruptive both to the operations of Canadian Copper and to the relationship between the company and the community.

As a source of supplies for the Canadian Copper Company, too, Sudbury was found wanting. In 1889, for example, General Manager E.D. Peters complained to company officials in Cleveland about the difficulties he was having in getting goods from local merchants. Orders would be placed, he explained, and then a couple of weeks later he would suddenly be told that the item was on back order or was not available. He had instructed his managers to order two months in advance, but this was obviously not always possible.[26] Peters also claimed that this inability of the local suppliers to meet the needs of the company resulted in the excessively high freight charges about which Mcintosh was querying him. Mcintosh's response was to suggest that the CCC appoint a company agent in Toronto who would order supplies from there, a proposal with which Peters readily agreed.[27] Over the course of the 1890s Sudbury's capacity as a supplier improved somewhat, but by 1901 the great volume of Canadian Copper's materials still came from southern Ontario businesses.[28]

The same situation applied in banking. Until January 1901, when the Bank of Toronto opened a Copper Cliff branch, Canadian Copper dealt locally with the Ontario Bank in Sudbury. According to Frederick Bernhard, however, this was little more than a formality, as the company's head office account was with the Bank of Toronto in Toronto. Therefore the number of transactions in Sudbury was quite small.[29]

All this is not intended to suggest that Canadian Copper looked upon the existence of Sudbury as completely irrelevant. The CCC certainly recognized that the community provided vital services in areas such as health (although at the end of the 1890s it eliminated this particular dependency by having the company doctor and medical facilities moved to Copper Cliff).[30] It also valued Sudbury's role as the commercial entrepot for the population of the entire region, including its employees.[31] This relieved it of the necessity of providing more than a minimal range of retail services in the shantytowns that congregated haphazardly around its mine and smelter sites. This meant too that formally, at least, it could banish from its properties those establishments it found undesirable, such as bars and taverns, knowing that they were available only a few miles away. In this respect, Sudbury was a useful adjunct indeed.

Still, the dependency of Canadian Copper on Sudbury in the 1890s was limited, a fact that in a general way accounted for the CCC's indifference to the community. The leading economic role played by Canadian Copper in the Sudbury district, the shift of its local headquarters to, and its growing identification with, Copper Cliff, its absentee ownership and executive direction, and the limited entrepreneurial and operational resources offered by the town all help to explain this attitude, but they do not adequately account for the genuine distrust that the company displayed towards the town. Its origins lay in a unique set of circumstances pertaining to the corporate ownership of the company, and related particularly to the activities of the one man who had been most responsible for the creation of the CCC in the first place, Samuel J. Ritchie.

SAMUEL J. RITCHIE

Between 1880 and 1885, Sam Ritchie, a Talmadge, Ohio, entrepreneur with a varied career in such things as carriage making and sewer construction, had invested and lost a fortune in iron mines and in the Central Ontario Railway in the Trenton, Ontario, area. Faced with personal bankruptcy, he cast about in the summer of 1885 for brighter prospects, and fortuitously stumbled across the copper-bearing properties in the Sudbury area. With the financial assistance of his Ohio associates, Thomas Cornell, Henry Payne and Stevenson Burke (to whom he was already heavily indebted), Ritchie purchased the most promising of these and conveyed them to the Canadian Copper Company for slightly over half of the $2 million worth of the CCC's original shares.[32]

The Canadian Copper Company thus began essentially as a Ritchie enterprise and at the outset his stamp was clearly imprinted on it. He became the company's first president, responsible for the start-up of mining operations in 1886. This included appointing the CCC's original management personnel. His brother-in-law, Lewis Ashmun – an Ohio farmer without mining experience – served as the first general manager and oversaw the company's development from 1886 to 1888. Its chemist from 1886 to 1891 was Francis Sperry, the son of one of Ritchie's former Akron partners in the carriage-making business. Ritchie drew on his businesses in Hastings County for other mining and managerial expertise. The CCC's first engineer and surveyor was John D. Evans of the Central Ontario Railway (COR), who travelled back and forth between Sudbury and Trenton until 1888 when he was permanently put on the Canadian Copper payroll because the COR could no longer afford his services. The company's first office manager, Tom Kirkwood, also came from the Hastings County area, as did several of the mining captains.[33] Most importantly, perhaps, Sudbury's first and for a long time only newspaper, the *Journal,* was founded in 1891 by J.A. Orr, formerly of Trenton, who later described himself and Ritchie as "very old friends."[34] Through people such as these, who during the first five years of the CCC's existence lived and worked in Sudbury, the Ritchie influence made itself felt not only on the company but in the town itself.

Ritchie's character was also a factor in the early rapport between the company and the town. He had a curious Jekyll-and-Hyde personality. His dealings with those to whom he related on relatively equal terms, such as his original COR railway partners, the McMullens, his financial associates such as Cornell, Payne and Burke, his American annexationist/business friends such as Benjamin Butterworth and Erastus Wiman, and even his legal advisors of whom there was a whole series, invariably ended bitterly in the law courts, after years of close and friendly relations.

With respect to the average Sudburian, on the other hand, Ritchie's behaviour was very different. His visits to the community, while seldom lengthy, were frequent. His extroverted, optimistic, "hail-fellow-well-met" approach to life was refreshing, particularly in small doses, and inspired long-lasting friendships both within the ranks of the Canadian Copper Company and the Sudbury community generally. Many years after the fact, Frederick Bernhard recalled that a number of Sudburians had told him "that Ritchie was one of the most likeable men they had ever met."[35] Certainly, whether it was prospector Thomas Frood, hotelier James McCormick, local mining entrepreneur

Aeneas McCharles, or CCC office manager and later roast-yard contractor Tom Kirkwood, Ritchie was accorded genuine respect and loyalty that persisted long after his formal connection with the community had ended.[36] During the 1880s, then, Ritchie personally contributed to the development of friendly relations between the company and the town.

Ritchie's success was clearly manifested in the presentation to him of a memorial from the community by McKim Township Reeve Stephen Fournier on 15 June 1891. Read by Fournier at the White House Hotel, against the backdrop, in Ritchie's words, of "nearly the whole town ... [complete] with a brass band,"[37] the memorial stated:

> We feel, sir, that to your perspicacity, energies and perseverence in developing the mining industry, of which you are the pioneer in this District, Sudbury owes to you what it is today. When you first came here, there were only a few buildings and a very small population, now we have a comparatively prosperous village with a population verging on two thousand.[38]

Ritchie also recalled (although this aspect of the event did not find its way into the account of the *Sudbury Journal*)[39] that the crowd later accompanied him to the railway station, insisting that the name of Sudbury be changed to "Ritchie," an idea he rejected.

Whether or not Ritchie entirely deserved the fulsome praise heaped upon him that June Monday is debatable; his reputation certainly owed much to his own very considerable talents for self-promotion. Still, justified or not, the fact that the memorial was presented at all was a measure of his personal popularity with Sudburians. Just as importantly, it was also a measure of the growing unpopularity of the Canadian Copper Company in the community, because by June of 1891 the CCC had clearly fallen into local disfavour. The timing of the presentation of the memorial was not at all accidental. It was in reality Sudbury's response to a formal statement issued by Canadian Copper three months earlier, on 16 March 1891, declaring that "Mr. S.J. Ritchie, formerly an officer of the Canadian Copper Company, has ceased to have any connection with the Company as a director, officer or stockholder and he does not represent the company in any capacity whatever."[40]

Sam Ritchie's fall from grace was not as sudden as the statement implied. Three and a half years earlier, in the autumn of 1887, he had been forced to resign both as president and director of Canadian Copper because he no longer held unencumbered title to a single company share.[41] Ritchie's difficulty was that by then, by his own accounting, he owed various creditors (but mostly Cornell, Payne and Burke) in excess of three-quarters of a million dollars. They in turn held as collateral his paper assets (Central Ontario Railway stocks, bonds and coupons, Canadian Copper Company stocks, and so forth) which had a face value of nearly $3 million, but a real worth of only a fraction of that amount.[42] For all intents and purposes, Ritchie was bankrupt. The straw that broke the camel's back in the fall of 1887 was the decision of the McMullen brothers, his former partners in the Central Ontario Railway, to sue him for failure to carry out an agreement to purchase their remaining COR bonds.[43] Both Ritchie and his Ohio backers began to fear that the assets of the CCC might become subject to seizure-hence the decision to sever all of Ritchie's legal ties with the company.[44]

For the next three and a half years, publicly and privately, Ritchie continued to act as though he were still the majority owner of the Canadian Copper Company. He was able to do so because Cornell, Payne and Burke carried him financially, extending him further credit when necessary, and covering his many notes when they fell due. It was in their own interests to do so, of course, in order to avoid further threats such as that posed by the McMullen lawsuit. At the same time, in the late 1880s they did get in return valuable services from him. Ritchie was nothing if not an adept travelling salesman, and these talents were put to good use. In 1889, for example, he went to Europe partly on a mission to find new nickel markets, but partly also to determine the market worth of the company itself.[45] In 1889-90, too, Ritchie used his contacts in high Washington places to get the American Congress to put nickel ore and matte on the free-duty list, as well as to interest the U.S. secretary of the navy, B.F. Tracy, in the armaments potential of nickel, an initiative that eventually led in 1891 to the first large-scale sale of the company's product.[46] Indeed, in September 1890, Ritchie was even reappointed director of the CCC with authorization to negotiate the terms of this sale.[47]

Had Ritchie only devoted his energies to Canadian Copper, his relationship with Cornell, Payne and Burke might have remained amicable. During the late 1880s, however, in his capacity as president of the struggling Central Ontario Railway, he continued to seek ways of salvaging its fortunes, and his own. The task was not an easy one. As he explained to an associate in October 1888, "nothing whatever can be done with this property [the COR and the Hastings County iron lands] alone. Some combination must be made by which it can be floated out by the aid of something else having more merit and this is what I have all along been trying to do."[48] The "something else having more merit," he decided, was the Canadian Copper Company, and for the next two and a half years he worked to amalgamate the COR and the CCC into a single company, to extend the COR from Hastings County to Sudbury, and to establish a nickel-steel manufacturing plant either at Trenton or Toronto.[49]

The plan was ambitious, even visionary. The problem was that Cornell, Payne and Burke were adamantly opposed to it. They were not prepared to risk the real worth of the Canadian Copper Company for the dubious prospect of saving the Central Ontario

Railway. Matters came to a head in early March 1891 when Ritchie, after repeatedly ignoring warnings not to commit Canadian Copper to the course of action he was proposing,[50] went one step too far and publicly challenged the authority of Canadian Copper management to enter into long-term sales agreements which conflicted with his own plans.[51] Faced with this challenge, the CCC directors finally concluded that they had no option but to disavow publicly the man popularly considered the "father" of Sudbury.

Sudburians were astounded by this turn of events, partly because of Ritchie's personal standing, but also because his proposal for an integrated nickel-steel industry in Ontario was extremely attractive to the community. The June 1891 memorial to Ritchie pointedly thanked him "specially for your untiring efforts, to have the Central Ontario Railway extended to this district" and expressed the hope that "God will be pleased to crown with success your great and noble enterprises to which you are devoting all your talents, and the most precious years of your life."[52] The lines of division between Ritchie and the community on the one side and the Cornell/Payne/Burke-controlled Canadian Copper Company on the other were clearly drawn.

Undoubtedly because he had personally been feuding with Ritchie since 1888, H.P. Mcintosh was particularly vigilant in his efforts to protect Canadian Copper from the former president.[53] Anyone in the slightest suspected of betraying the company's interests to Ritchie was not long tolerated. Francis Sperry, the CCC chemist since the company's inception, left its employ to become an independent mining consultant in Sudbury when the break with Ritchie occurred. Another casualty was John Evans, the CCC general manager from 1890 to 1893. As a longtime acquaintance of Ritchie, he was required to give evidence in one of Ritchie's many lawsuits against his former associates in early May 1893. On 22 June he was summarily dismissed, ostensibly for reasons of financial constraint, but in his own mind at least (and probably for good reason) because of the testimony he had given.[54]

The rupture between Canadian Copper and the Sudbury community persisted for the remainder of the 1890s, exacerbated by Ritchie's ouster from control of the Central Ontario Railway in 1892. In an attempt to re-establish his claims of ownership to both the COR and the CCC, Ritchie initiated a series of legal suits, responded to in kind by the various CCC parties or their estates. Thanks to James Orr and the *Sudbury Journal,* Sudburians were able to follow his efforts in the courts of Ontario and Ohio through the mid-1890s.[55] Some were quick to assure Ritchie of their support. Tom Kirkwood, for example, informed him in 1893 that "you will see that you still have friends here, and that this is not a H.P. Mcintosh town, the feeling here is for you and we all hope that you will succeed in your suits with the Company."[56] The *Journal,* though, was notably silent about the ultimate failure of Ritchie to regain control of the CCC when, in November 1897, the assets he had assigned as collateral were sold by judicial auction in Cleveland, purchased from themselves at a previously agreed-upon price by their holders.[57]

The legal confirmation of Burke et al.'s ownership of Canadian Copper in 1897 in no way diminished Ritchie's zeal. Determined to get revenge if he could not get retribution, he launched a three-pronged attack on the Canadian Copper Company. First of all, he commenced a new round of lawsuits in the courts against the company calling for its dissolution on a variety of grounds. One was its failure to live up to the promise Ritchie claimed had been made to the Parliament of Canada at the time of its naturalization in 1886 to refine its matte within Canada.[58] Another was premised on the contention that Canadian Copper had no legal right to operate in Ontario because it held a federal charter and natural resources were under provincial jurisdiction.[59]

Secondly, on the business front Ritchie found a new set of allies, Hamilton industrialists John Patterson, A.T. Wood and John M. Gibson, and proceeded to promote a series of new nickel-steel-related companies – the Nickel Copper Company, the Hoepfner Refining Company and the Nickel Steel Company of Canada – to compete with Canadian Copper and its Orford refining partner. While these ventures ultimately failed, Ritchie and his associates were able publicly to maintain the appearance of competitive viability and put pressure on the CCC for four years.[60]

The third prong of Ritchie's attack on Canadian Copper was political in nature. Once an advocate of the annexation of Canada to the United States and a virulent opponent of any measures which restricted trade between the two countries, he experienced in the late 1890s a radical conversion to Canadian economic nationalism.[61] In order to facilitate the development of the nickel-steel industry in Canada, Ritchie spearheaded a campaign to persuade the federal government of Sir Wilfrid Laurier to impose an export duty on nickel and copper ores and mattes. For this campaign he had chosen his business partners well, as Patterson, Wood and Gibson were prominent and influential Liberals. In 1897 the Parliament of Canada actually passed legislation providing for such duties, but appended a proviso requiring the Laurier government to adopt an order-in-council to put them into effect.[62]

When the Laurier government failed to implement the export duties, Ritchie and his associates redirected their energies in 1899 towards the provincial government of Ontario, calling upon it to impose a "manufacturing condition" on nickel ore similar to that applied to saw logs in 1897. Here they were on even firmer political ground, as Gibson was a member of that Liberal administration, having served first as commissioner of crown lands in the Hardy cabinet before his appointment to the position of attorney general by Premier Ross.[63] In addition, the director of the Bureau of Mines, Archibald Blue, whose friendship with Ritchie dated from 1886, was actively working behind the scenes to force the government's hand.[64] As a result, the Mines Act was amended in 1900 to provide for the imposition of a provincial licence fee on unrefined matte destined for export, although once again an order-in-council had to be issued by the provincial cabinet for this to become operative and none was forthcoming.[65]

Ritchie's entrance into the political forum with the support of influential men like John M. Gibson dramatically changed the nature of the conflict from the perspective of the Canadian Copper Company. So long as the battles were being fought in the corporate boardrooms or the law courts of Ontario and Ohio, Canadian Copper was well able to take care of itself. Once the venue shifted to the political arena, the company was in a much more difficult position. By 1898 the CCC genuinely needed a strong show of popular support from Sudbury and its citizens.

IN SEARCH OF HARMONY

Getting that public demonstration of support was not an easy task. With respect to the export duty on nickel matte, for example, James Orr claimed in the *Sudbury Journal* on 24 February 1898 that "with the exception of one company, the people of this district, where the nickel mining industry is situated, are almost unanimously in favor of such action on the part of the Government, for the benefit of the country at large, and in order to promote the development of our nickel mines."[66] Canadian Copper set about to prove Orr wrong. On 25 February Mcintosh instructed the CCC general manager, James McArthur, to arrange a public meeting in Sudbury to prepare a memorial against the duty, as well as to circulate petitions to be signed to the same end, "all this to be done to prevent it being known such work originated with us."[67]

McArthur agreed to make the arrangements as best he could but warned Mcintosh that it might be expensive:

> To accomplish this it will be necessary to do the work thro' the Sudbury papers and the Mayor Mr. Qochrane]. and this will require money. The latter's time is worth something and the papers, as you know will not publish anything in our interest unless it is made to their interest. If necessary we thought we might go as high as 100.00 [dollars] with Mr. C. and whatever else is necessary for the others. Of course it will be understood in any event that the funds are coming from private parties to prevent the export duty in the interests of such contributing individuals. Mr. Cochrane will be seen this afternoon by Mr. Glidden and another party who can influence the *"Journal,"* it is expected. Mr. Cochrane will likely do what is right with "News."[68]

McArthur proved to be overly sanguine with respect to the *Sudbury Journal,* which was not at this point prepared to drop its campaign for an export duty on nickel, but both the mayor, Frank Cochrane, and the *Sudbury Mining News* did what was asked of them.[69] Not only was a resolution against an export duty passed by town council (a fact which the *Journal* attributed to the absence of the pro-duty aldermen), but a delegation of Sudburians was despatched to Ottawa to meet with the Laurier government.[70] Even Aeneas McCharles had to admit that the CCC's efforts were effective, noting that although "the majority of the people of this district are in favor of an export duty on nickel ores and mattes, ... nevertheless the most of them are actively signing a petition the company is getting up against it."[71]

The CCC's orchestration of popular opinion in Sudbury against an export duty on nickel may have had the desired effect on the Laurier government – the enabling order-in-council was never issued – but the strategy did not necessarily win it genuine friends. This, certainly, was the view of Frederick Bernhard, who was transferred from Cleveland in September 1899. As an example of the anti-company feeling he found, Bernhard cited the "evident glee" with which the Methodist minister in Sudbury privately predicted that the newly discovered metal, aluminum, would soon drive nickel out of the marketplace.[72] Bernhard was also present at the time of the *Journal's* most blistering attack on Stevenson Burke specifically and on the Canadian Copper Company generally for the impact of sulphur-fume emissions on the Sudbury region. Incensed by a report that Burke was threatening to take legal action against the Pennsylvania Railroad for smoke it was emitting onto his property in Cleveland, Orr railed against the devastation wreaked throughout the Sudbury district by the CCC's heap-roasting process.[73]

Barely a month later, in November 1899, Canadian Copper initiated a campaign to improve relations with Sudbury through the appointment of a new assistant general manager, A.P. Turner. Unlike his predecessors, who had come to the company from outside, Turner had worked for Canadian Copper in Cleveland for six years. He brought with him not only a vital familiarity with the company's operations, but also the confidence of H.P. Mcintosh, whose position as president of the Guardian Trust Company in Cleveland was increasingly occupying his time. According to Frederick Bernhard, Turner moved almost immediately upon his arrival to refurbish the company's image in Sudbury, making a special effort to socialize with the leading businessmen of the community.[74]

At the top of Turner's list was winning over James Orr and the *Sudbury Journal.* On 1 February 1900 he reported at length to Mcintosh on his meeting with Orr. Turner's pitch was a clever one, as he began by telling Orr that the *Journal's* defence of Ritchie and attacks on Canadian Copper did not bother the people in Cleveland but they were "embarrassing for me

[Turner], inasmuch as I was going to live here." He then pointed out to Orr that he could "give him more or less business, but that he must understand that I could not do so consistently if he continued the publication of such articles." Orr immediately apologized and promised that Turner "would have no further complaint in this direction." Turner added that Orr "even went so far as to say that he had begun to feel that possibly his position in regard to the export duty was not warranted by the present conditions of the industry." In closing, Orr asked that Turner from time to time supply the *Journal* with information about the operations of Canadian Copper, a request to which Turner agreed. "Of course," he wrote Mcintosh, "nothing should be told him except what people generally know or could find out from our reports to the Government."[75]

Turner's appeal obviously had the desired effect. Over the next twelve months the *Journal* edged its way first to a position of neutrality and then to the outright defence of the CCC.[76] By January 1901, Turner was actually writing unsigned editorials for the paper against the continuing efforts of Sam Ritchie to persuade the government of Canada to activate its export duty and the Ontario government to enforce the manufacturing condition on nickel. The cost to Canadian Copper was not particularly high: only the purchase of 200 copies of the edition as well as occasionally supplying further information to Orr about its operations for publication.[77] The price to Orr was the embarrassment of explaining his about-face, a reversal that had quickly been noted in the Toronto press. His somewhat lame explanation was that the appearance of the Mond interests from Great Britain as competition to the CCC, and the continued failure of the Ritchie-Patterson group to establish viable operations, had totally altered the situation from only a year and a half before.[78]

By 1902 Canadian Copper had accomplished what it had set out to do. On all fronts the Ritchie challenge was beaten back. In the political field the orders-in-council Ritchie had so desperately sought remained unissued, the Liberal governments at both the federal and provincial levels having been persuaded of the validity of the CCC case. His business alliance with the Hamilton magnates similarly failed to yield results, mainly because the oft-promised nickel-refining technology did not materialize. Finally, Ritchie's latest round of lawsuits went the way of his earlier ones, and in March of 1902 he was persuaded to withdraw all of them, facilitating the sale of Canadian Copper to the International Nickel Company.[79]

With the final removal of the Ritchie factor from the equation, the Canadian Copper's relationship with Sudbury took a turn for the better. According to Frederick Bernhard, A.P. Turner worked very hard in the early 1900s to maintain the more harmonious relations that he had so assiduously cultivated.[80] Still, this second era of company-town goodwill that closed the Cleveland phase of the Canadian Copper Company was very different from that of the 1880s. More orchestrated, artificial, motivated by corporate self-interest, it was also short-lived. Again according to Bernhard, as soon as Turner was appointed president of Canadian Copper after its transfer to Inco in 1902, his conciliatory attitude towards Sudbury gave way to a "rather arbitrary" approach that quickly dissipated the goodwill he had built up.[81]

CONCLUSION

The relationship that developed between Sudbury and Canadian Copper during its formative years has been described as symbiotic.[82] From the perspective of the company, however, this was true only to a limited extent, for while there was a degree of mutual interdependency, Sudbury's reliance upon Canadian Copper was very much greater than the reverse. The imbalance permitted Canadian Copper the luxury of dealing with Sudbury pretty much as it wanted. It was largely indifferent to the town's interests and opinions until unique circumstances and individual personalities created a situation in which it temporarily needed Sudbury's support. Factors such as the CCC's dominant economic position, absentee ownership and remote executive direction fostered a general attitude of detachment and disinterest that was altered only by the particular exigencies of the Ritchie-versus-Cornell/Payne/Burke feud. The next stage of the company-town association would lack this unpredictable dimension. As an Inco subsidiary, Canadian

Copper's relations with Sudbury would be shaped directly by more impersonal corporate and managerial considerations.

Notes

Author's note: The author is indebted to Ron Orasi and the Inco Archives for assistance in the preparation of this article.

[1] Robert Robson, *Canadian Single Industry Communities: A Literature Review and Annotated Bibliography* (Sackville, N.B., 1986), 1-26.

[2] G.A. Stelter, "The Origins of a Company Town: Sudbury in the Nineteenth Century," *Laurentian University Review* 3, no.1 (February 1971); G.A. Stelter, "Community Development in Toronto's Commercial Empire: The Industrial Towns of the Nickel Belt, 1883-1931," *Laurentian University Review* 6, no.3 (June 1974); Noel Beach, "Nickel Capital: Sudbury and the Nickel Industry, 1905-1925," *Laurentian University Review* 6, no.3 (June 1974); Eileen Goltz, "Genesis and Growth of a Company Town: Copper Cliff, 1886-1920" (M.A. thesis, Laurentian University, Sudbury, 1983).

[3] Goltz, "Genesis and Growth of a Company Town," 186-91.

[4] Stelter, "The Origins of a Company Town," 18-28.

[5] *Sudbury journal,* editorial, "Consistency, Etc.," 5 October 1899.

[6] Inco Archives, Sudbury [hereafter abbreviated as IAS], Microfilmed Ritchie Papers, Ritchie to Lewis Ashmun, 19 December 1886.

[7] Stelter, "Community Development," 9-10; Goltz, "Genesis and Growth of a Company Town," 46-47.

[8] IAS, Letter Book No.5, A.P. Turner to Mcintosh, 18 July 1901.

[9] Thomas Cornell was president of the City National Bank of Akron. Stevenson Burke, a corporation lawyer, was president of several railways in Ohio, including the Cleveland, Columbus, Cincinnati and Indianapolis Railway Company, the Cleveland and Mahoning Valley Railway Company, the Indianapolis and St. Louis Railway Company and the Columbus, Hocking Valley and Toledo Railway Company, as well as of the Snow Fork and Cleveland Coal Company. Henry B. Payne, also a lawyer by training, made his fortune from railways such as the Cleveland and Columbus, and the Cleveland, Painesville and Ashtabula. Payne later diversified his interests, investing in the Citizens' Savings and Loan Association and the Union Steel Screw Company, both of Cleveland. See S.P. Orth, *History of Cleveland – Biographical* (Cleveland, n.d.), 16-30, 59-67; W.G. Ross, *Cleveland: The Making of a City* (Cleveland, 1950), 443-4, 492-97.

[10] E.A. Collins, "Early Days of the International Nickel Company" (Paper presented to the An-

nual General Meeting of the International Nickel Company of Canada, April 1933); reprinted in *The Canadian Institute of Mining and Metallurgy* (1933), 428-52.

[11] Mcintosh once reprimanded the CCC general manager, Dr. E.D. Peters, for the poor quality of his handwritten reports. "Allow us to suggest," he wrote in August 1889, "that, in writing your letters, you do not crowd the matter so, nor use the margins, nor invert the sheets, for it makes it very difficult to read after they are filed. Please write them all out full hand, and we will pay for the paper" (cited in Collins, "Early Days," 449-50).

[12] IAS, Ritchie Papers, RS1350 Box E, Mcintosh to Sam Ritchie, 20 August 1888.

[13] IAS, Letter Book No.1, E.D. Peters to Mcintosh, 29 November 1889.

[14] IAS, Letter Book No.2, Evans to Mcintosh, 25 January 1892.

[15] IAS, Letter Book No.2, Evans to Mcintosh, 27 July 1891.

[16] IAS, Ritchie Papers, 929-53, Sperry to S.J. Ritchie, 24 June 1891.

[17] See, for example, IAS, Diary of John D. Evans, 23 May 1891, 2 June 1891, 17 June 1891, 10 October 1891, 16 February 1893.

[18] James A. Orr, "Profits of Nickel Mining," *Sudbury Journal,* 16 January 1896; and "Why Is It?" *Sudbury Journal,* 20 February 1896.

[19] IAS, Letter Book No.2, Glidden to Mcintosh, 17 February 1898.

[20] O.W. Main, *The Canadian Nickel Industry* (Toronto, 1955), chaps. 2-4, 18-60.

[21] The daily correspondence between John D. Evans and Mcintosh, 1890-93, often included reports on the company's competitors. See, for example, IAS, Letter Book No.2, Evans to Mcintosh, 18 April 1892.

[22] IAS, 927-55, The Canadian Copper Company, Preliminary Report for year ending 31 December 1896; "Is There a Combine?" *Sudbury Journal,* 30 January 1896.

[23] IAS, Frederick P. Bernhard, "Historical Notes on INCO" (unpublished manuscript, 1953), 23.

[24] IAS, Evans Diary, 22 June 1893.

[25] Ibid., dates as noted.

[26] IAS, Letter Book No.1, Peters to Mcintosh, 20 October 1889.

[27] Ibid., 22 October 1889.

[28] IAS, Letter Book No.5, Turner to Mcintosh, 11 January 1901. This letter includes a list of thirty-nine companies with which the IAS was dealing, none of which was in Sudbury, and only two of which were in Northern Ontario, both at Warren.

[29] IAS, Bernhard, "Historical Notes on INCO," 46.

[30] IAS, Letter Book No.5, Turner to Mcintosh, 22 September 1900.

[31] See Goltz, "Genesis and Growth of a Company Town," 89-92.

[32] John F. Thompson and Norman Beasley, *For the Years to Come* (Toronto, 1960), 16-38.

[33] IAS, Ritchie Papers, 929-53, L.H. Ashmun to Ritchie, 24 April 1886. (See also Collins, "Early Days," 444.)

[34] IAS, Letter Book No.5, A.P. Turner to Mcintosh, 1 February 1900.

[35] IAS, Bernhard, "Historical Notes on INCO," 6.

[36] IAS, Ritchie Papers, RS1350 Box E, Thomas Frood to Ritchie, 21 May 1886; McCharles to Ritchie, 23 June 1899; Kirkwood to Ritchie, 16 June 1902.

[37] IAS, Ritchie Papers, 929-53, Ritchie to Clara Belle Ritchie, 16 June 1891.

[38] IAS, Ritchie Papers, 929-53, Memorial to S.J. Ritchie, 15 June 1891.

[39] IAS, Ritchie Papers, 929-53, Ritchie to Clara Belle Ritchie, 16 June 1891; *Sudbury Journal,* "Address to Mr. Ritchie," 18 June 1891.

[40] IAS, *925-AA,* Canadian Copper Company Circular Letter, 16 March 1891.

[41] Western Reserve Historical Society, Cleveland, Ritchie Papers, Box 19, draft of a letter from Ritchie to Mcintosh, 27 August 1887. The resignation was not accepted until 13 September 1887. See Thompson and Beasley, *For the Years to Come,* 53. Thompson and Beasley claim that on the same date Ritchie

resigned as director, but company records indicate that he gave up this position as of 18 December 1887.

[42] IAS, Microfilmed Ritchie Papers, Ritchie's Statement of Assets and Liabilities, 10 November 1887.

[43] IAS, Microfilmed Ritchie Papers, Ritchie to Messrs. Bell and Biggar, Belleville, 8 November 1887.

[44] IAS, Microfilmed Ritchie Papers, Mcintosh to Ritchie, 21 September 1887.

[45] See Philip Smith, *Harvest from the Rock* (Toronto, 1986), 80-81.

[46] Stelter, "The Origins of a Company Town," 15-17. See also IAS, Microfilmed Ritchie Papers, Ritchie to Sir Charles Tupper, 30 April 1890; and Microfilmed Ritchie Papers, Ritchie's Statement of the History of His Businesses, undated, c. April 1892.

[47] IAS, 932-23, Canadian Copper Company, Minute Book No.1, 8, 27 September 1890.

[48] IAS, Ritchie Papers, 929-53, Ritchie to William Coe, 7 October 1888.

[49] See, for example, Public Archives of Canada, Macdonald Papers, vol.257, 116384-87, Ritchie to Macdonald, 19 October 1890.

[50] See, for example, the Western Reserve Historical Society, Cleveland, Ritchie Papers, Box No.5, Stevenson Burke to Ritchie, 16 January 1891; and IAS, Ritchie Papers, 929-

53, Burke to Ritchie, 24 January 1891. See also G.A. Stelter, "The Origins of a Company Town," 16-17.

[51] IAS, Ritchie Papers, 929-53, Ritchie to Robert M. Thompson, 10 March 1891.

[52] *Sudbury Journal,* "Address to Mr. Ritchie," 18 June 1891.

[53] The split between Mcintosh and Ritchie occurred in August 1888, when Ritchie accused Mcintosh of undermining his standing with Payne et al. See IAS, Ritchie Papers, RS1350 Box E, Mcintosh to Ritchie, 8 August 1888.

[54] IAS, Evans Diary, 8 May 1893; 22 June 1893.

[55] See, for example, *Sudbury Journal,* 2 March 1893; 9 March 1893; 11 May 1893; 24 August 1893; 5 October 1893; 25 October 1894; 17 January 1895; 1 April 1897.

[56] IAS, Ritchie Papers, RS1350 Box E, Kirkwood to Ritchie, 6 March 1893.

[57] IAS, 925-44, Report by Irvin Belford, special commissioner, Circuit Court of the U.S. for Northern District of Ohio, Eastern Division, 6 November 1897.

[58] Smith, *Harvest from the Rock,* 110.

[59] IAS, Microfilmed Ritchie Papers, Ritchie to Judge Upson, 16 July 1899.

[60] H.V. Nelles, *The Politics of Development* (Toronto, 1975), 90; Thompson and Beaseley, *For the Years to Come,* 116.

[61] Re: Ritchie's annexationist sentiments, see, for example, IAS, Ritchie Papers, 929-53, Ritchie to Erastus Wiman, 27 February 1887. As late as June 1893, he was still sympathetic to the cause. See IAS, Ritchie Papers, RS1350 Box E, Edward Farrer to Ritchie, 2 June 1893. For his late 1890s views see S.J. Ritchie, *The Question of Export Duties on Nickel and Copper Mattes: The Anonymous Circular Answered* (Ottawa, 1898).

[62] Nelles, *The Politics of Development,* 90-91.

[63] Ibid.

[64] IAS, Evans Diary, 10 June 1886. Indeed, Blue had strongly advised Ritchie to work with John Patterson and the Hamilton people. IAS, Ritchie Papers, 929-53, Archibald Blue to Ritchie, 7 January 1898. See also Blue to Ritchie, 17 March 1899.

[65] Smith, *Harvest from the Rock,* 112.

[66] *Sudbury Journal,* "An Export Duty on Nickel," 24 February 1898.

[67] IAS, Letter Book No.2, McArthur to Mcintosh, 26 February 1898.

[68] Ibid.

[69] *Sudbury Journal,* "A Dastardly Letter," 7 April 1898. This article dealt with a letter by Frank Cochrane in the *Sudbury Mining News* which pleaded the case against an export duty on nickel.

[70] *Sudbury Journal,* "The New Caledonia Mines," 24 March 1898; IAS, General Letter Book No.2, Glidden to Mcintosh, 10 March 1898.

[71] IAS, Ritchie Papers, RS1350, Box E, McCharles to Ritchie, 8 March 1899.

[72] IAS, Bernhard, "Historical Notes on INCO," 29.

[73] *Sudbury Journal,* "*To* Get Rid of Smoke," 5 October 1899; *Sudbury Journal,* editorial, "Consistency, Etc.," 5 October 1899. See also Stelter, "The Origins of a Company Town," 28-29.

[74] IAS, Bernhard, "Historical Notes on INCO," 18.

[75] IAS, Letter Book No.5, Turner to Mcintosh, 1 February 1900.

[76] See, for example, the *Sudbury Journal,* "Big Deputation," 26 April 1900; 26 July 1900; 20 September 1900.

[77] IAS, Letter Book No.5, Turner to Mcintosh, 18 and 25 January 1901.

[78] *Sudbury Journal,* "The Nickel Refining Question," 23 February 1901.

[79] IAS, Microfilmed Ritchie Papers, Ritchie to George Burn, 13 March 1902. The details of Ritchie's arrangement with Burke and the estates of Cornell and Payne are not clear, but within a year of the formation of Inco, Ritchie miraculously resurfaced as president and ma-

jority owner of the still-struggling Central Ontario Railway.

[80] IAS, Bernhard, "Historical Notes on INCO," 18.

[81] Ibid.

[82] Goltz, "The Genesis and Growth of a Company Town," 89-92.

Chapter 3

All That Glitters: Speculation and Development in the "Minor" Minerals of the Sudbury Area

Peter V. Krats

A historical study of the Sudbury area reveals much about the relationship between the settled south and the frontier north, especially about the impact of outside corporations on local staples exploitation. Yet individual initiatives, inspired by resolutely optimistic assessments of the resource base, were also crucial to regional development. The relentless enthusiasms of Samuel Ritchie are fairly well known; not so well known, however, is the work of mine promoters calling attention to minerals other than nickel. Their efforts, though rarely successful, reveal rarely explored facets of the Sudbury area experience.

Beginning with the copper boom of 1883, this paper examines the continuing fascination with a variety of minerals – gold, lead, iron, even diamonds and coal – in the Sudbury area, and provides information on regional infrastructure – roads, services, jobs and the like – that is not attributable to the nickel industry. At the same time, the fascinating characters of the mineral "boosters" cannot be ignored: cheats and frauds abounded along with men who had an abiding faith in the local deposits. Honest boosterism and fraud, of course, can follow closely aligned convoluted paths, and both add much "colour" to the local setting. More importantly, the minor mineral industry, along with the often-ignored renewable resource sector, created a varied economy belying the Sudbury area's reputation as a monopoly-directed "Nickel Capital."[1]

"COPPER CAPITAL"

The entry of the Canadian Pacific Railway crews into what would become known as the Sudbury area initiated changes beyond the provision of transportation, communications and social services. The easily recognized signs of copper-stained rock outcrops strewn about the Sudbury area prompted rapid investigation.[2] Members of the CPR advance crew were first to move; one Sunday in August 1883, Francis Crean and John Flanagan located a deposit thought to contain copper. They were not alone in their curiosity: George Thompson applied for a "miners license to work a claim" in McKim Township on 8 February 1884; seventeen days later John Loughrin, Thomas and William Murray, and Harry Abbott applied for the deposit located by Crean and Flanagan.[3] With the application for the Murray location, the "rush" was on; as spring begat summer, favourable geological reports and optimistic publicity caused "blanket" claiming to run rampant.[4]

Copper, then, sparked the first local mineral boom, bringing much-needed income to an area hard hit by the departure of the railway crews. Practical mining promised even greater things to come. First to move was a "shrewd Yankee" who financed work on a copper deposit in Drury Township. The Crean-Harwood mine was optioned by Stewart Webb, linked by marriage to the wealthy Vanderbilts. Work began in May 1885, with mining engineer John Hurd supervising a force of twenty-two men who drove a shaft to a sixty-foot depth. Then Hurd made his decision: the "New York gentlemen" financing the mine "abandoned it and relinquished their lease on the advice of their manager."[5]

This failure did not blunt speculative enthusiasm, because by then ten or more "copper" locations had been improved.[6] Besides providing a valuable boost to the local economy, the publicity surrounding these improvements caught the attention of Samuel J. Ritchie. Dreaming on an extravagant scale, Ritchie and his Ohio backers acquired properties and opted to do some mining. The Canadian Copper Company was incorporated on 6 January 1886. As the name suggests, the intent was to mine copper: accordingly, Ritchie launched a relentless promotional campaign culminating in the fall 1886 visit of William Van Home and Sir George Stephen, soon followed by Sir Charles Tupper and Lady Macdonald.[7]

Exaggeration concerning the copper deposits grew exponentially as the regional, national and even the international press reported these visits. By late 1886 Sudbury was "in a perfect fever of mining excitement," with a floating and stationary population of up to 3,000 persons. Four rapidly constructed hotels and numerous other businesses, all charging excessive prices, were kept busy thanks to reports asserting the existence of "mountains" of copper. Sudbury, in short, experienced a full-fledged boom, and the copper fever also infected the new mining camps, especially Copper Cliff.[8]

But the copper fever broke with the discovery that Canadian Copper's ore averaged only 7 percent mineral content despite hand cobbing. Worse, the ore tests proved disastrous, revealing an ore containing not only copper but nickel, a combination that rendered standard refining processes ineffective. Besides, there was almost no market for nickel. By December 1886 the "tainted" nature of the ore was public knowledge and the copper frenzy no more than a bittersweet memory.[9]

With sharp decline in ore shipments in 1887, Canadian Copper faced ruin. Its only hope was to discover a cost-effective means of nickel-copper refining and to develop a market for nickel. But even as it launched itself on that well-known saga, mineral speculators and prospectors, with smaller investments in copper but also less capital, turned their attentions to minerals that had been all but forgotten during the glory days of copper. Surely, they insisted, the numerous hills about Sudbury offered minerals galore to those willing to take up the challenge of finding them.

MINERAL WEALTH ABOUNDS

The optimistic convictions and exploitative outlook of the mineral speculator were most evident in the search for gold, the merest hint of which stirred great excitement. Thus, the pessimism brought on by the news of nickel "tainting" was alleviated in September 1887 by Henri Ranger's discovery of gold ore in Denison Township. Ranger, Robert Tough and James Stobie received their patent for the Vermillion mine on 9 October and immediately undertook site preparation. They also recruited A.G. Duncan, an experienced mining engineer, who set out on a search for investors. Duncan interested Robert Hill, Volney W. Foster and several other wealthy Chicagoans, who along with B.E. Charlton of Hamilton and the patentees formed the Vermillion Mining Company. Encouraged when the mine yielded $8,000 in gold, they invested $70,000 in new equipment. By the summer of 1888 three shafts and a three-stamp mill kept about 40 men busy. The owners, in answer to criticisms over slow development and limited production, increased the workforce to 60 men and undertook further exploratory tests. When the tests revealed promising signs of platinum and nickel-copper ores as well as gold, they announced plans for a sixty-stamp mill, an enlarged plant employing over 200 men and even a company town-Ranger – which was to rise a little south of the mine alongside the CPR's Algoma Branch. Canadian Copper, which had been quietly acquiring Vermillion shares, had seen enough: the purchase of Robert Tough's stock gave it a controlling interest. As the nickel-copper firm needed no new ore, the mine was soon closed.[10]

The Vermillion mine's brief moment of glory inspired industry, scientific and newspaper coverage of the "Sudbury gold fields." Reputable individuals were caught up in the excitement: Thomas Johnson, assistant commissioner of crown lands, proclaimed the area "simply marvellous. It is one of the richest mining districts ever discovered. I had heard of its repute, but I did not believe it till I saw it for myself."[11] Such accolades spurred a gold rush that by April 1888 attracted 1,000 or more hopefuls to Sudbury. Prices for food, supplies, lodging and real estate inflated crazily as newspapers reported "Gold! Gold! Gold! The palmiest days of California are surpassed."[12] The reality, as Aeneas McCharles scornfully noted, was less appealing. The rush, he reported, was dominated by "city dudes and others who knew nothing practically about exploring, till you would think that Barnum's circus had got loose; but the majority of them went home as soon as the liquor ... gave out or the flies came." The "dudes" were too late in any case, for all through the winter "[speculators] from Toronto and Montreal ... bought up claims blindly for miles around the [Vermillion] mine, without knowing whether there was any mineral on them or not. And worse still, no effort has been made in most cases to find out ... the owners bought like fools and are waiting for bigger fools to buy them out."[13]

Most speculators and explorers deserved McCharles's scorn, but the gold excitement did result in forty or so small "finds," either by independent miners or new firms like the Sudbury Mining Company. Among the most prominent of the independents were the Simpson brothers of Cranston, Illinois, who from 1888 to 1891 worked seven small Graham Township pits.[14] Only a few sites, however, showed enough promise to warrant full-scale mining. The Creighton Gold Mining Company, financed out of Ottawa and Toronto, made perhaps the largest investment, sinking a 160-foot shaft and installing a three-storey gold mill that together cost $30,000. The operation, managed by J.R. Gordon, had employed as many as sixty men, but closed when the gold ore petered out late in 1893.[15]

The other major work inspired by the Vermillion rush was a joint venture of Aeneas McCharles and Robert H. Ahn, two entrepreneurs of indefatigable spirit. The former arrived at Whitefish early in 1889, attracted by reports of gold; Ahn, for his part, hoped to establish his metallurgical expertise by building a "wet process" mill for refining the royal and platinum group metals. For a brief moment these two self-assured men moved in unison: Ahn's Algoma Gold Mining and Reduction Company mill was to be the centrepiece of Nickel City, McCharles's self-proclaimed capital of the mineral range. Work on the picturesque site just east of Whitefish began in 1889, with a gridiron townsite laid down in 1890; construction of the gold mill began that fall. But the plant, plagued by technical failure (Ahn's critics cried fraud), closed in November 1891. Under the liquidation scheme of April 1892, plant employees received 75 percent of their wages and company investors suffered a loss of perhaps $13,000. Ahn fared better, moving on to manage other mining operations while waging a spirited war of words with his critics. McCharles, whose town promotion scheme lay in ruins, was an especially vehement foe; his Nickel City failure left him a bitter man despite better days to come.[16]

Notwithstanding the notoriety of Nickel City and the long list of failures, gold retained its allure, with prospectors drawn to the sometimes rich, albeit small, gold veins near Lake Wanapitei. Promotion in the Toronto press, encouraged by often unscrupulous mine brokers and local businessmen like Frank Cochrane, soon blossomed into a new rush. Investors moved to establish gold-mining companies and local businessmen stood poised to take advantage of the new opportunities: Dan O'Connor placed a small steamboat (*The Maid of the Mill*) on the lake and there was talk of a new lakeside town (Sawyer). Miners and labourers also found work as various pits were improved; the Crystal, Comstock and Bonanza were the largest. Most importantly, at least to the investors, a few thousand dollars in gold was processed through the new stamp mills.[17]

Gold production soon began to decline, forcing speculators ever farther afield. The search for gold soon reached the upper reaches of the Vermilion River where, in 1896, N.S. Jenkins located gold-bearing gravel beds in northeast Hanmer Township. Jenkins's work, financed by the Great Northern Mining and Development Company, rekindled speculative enthusiasm for Sudbury-area gold. The *Globe,* for instance, reported:

> Gold and rumours of gold. The Spanish-American war is of secondary importance to the good people of Sudbury ... for the moment at least their pride in their nickel mines has been supplanted by a far-reaching interest in a strip of land along the Vermilion River,... which is said to contain placer gold beds ... the people are sanguine that further search will reveal deposits of a richness and extent sufficient to cause a revival of the palmy days of the nickel boom.[18]

Dozens of prospecting teams headed north to the Vermilion River's upper reaches, but tantalizing gold recoveries proved insufficient to hold speculative interest in the face of gold booms in northwestern Ontario and especially the Klondike.[19]

Flakes of gold fired many, often short-lived, dreams, but true optimists saw fortunes to be made in many other "motherlodes." The list was a long and sometimes unlikely one, including silver, tin, molybdenum, cobalt, peat and even Balfour Township diamonds.[20] No deposit attracted more attention, however, than the "coal" seams of the Sudbury Basin's interior valley. Though sought for some time by mine brokers like J.R. Gordon and William McVittie, the first confirmed find was made in the spring of 1896. Late in October 1896, the Ontario Bureau of Mines despatched professor A.P. Coleman to study the by now well-known deposits; his journey triggered an avalanche of hysterical reports. A *Toronto World* headline trumpeted that "Ontario Has Coal Sure!"; the *Globe* and *Toronto Star* were equally enthusiastic. Coleman's conclusion that the deposits consisted of a small quantity of worthless anthracitic carbon – anthraxolite – hardly slowed the pace of speculation. Three or more firms were in the field by December 1896, sinking shafts and erecting mine buildings at several locations. The Citizens' Gold and Coal Company was most persistent, carrying on its work well into 1898 and thereby earning local plaudits for its "faith and pluck." But "pluck" did not produce markets for the poor-quality carbon, so the firm slowly withered into oblivion.[21]

Even in failure the futile search for coal had a positive impact, for the prospectors scouring the valley took renewed notice of the galena (lead sulphide) deposits first staked in 1888. Their interest, not surprisingly, drew investor curiosity. Buffalo and Toronto capitalists financed a diamond drill campaign beginning in November 1896; positive results saw the Balfour Mining Company commit a large sum – estimates ran to $30,000 – towards property development. But high operating costs, complex ores and weak lead markets took their toll: large-scale work stopped late in 1898, leaving a drowned shaft and a 300-ton ore pile as mute testimony to monies lost.[22]

AN ALTERNATIVE TO MONOPOLY

The demise of firm after firm and recurring losses of investments provided seemingly ample evidence that the minor minerals of the Sudbury area could not by themselves sustain practical mining operations. But the always treacherous path to nickel riches was all but closed following the formation of International Nickel. To be sure, Mond Nickel survived, thanks to its founders' science and fortune, but this was little solace to other entrepreneurs locked out by the market and financial control exercised by the international nickel cartel. Wise investors looked elsewhere, while those determined to exploit the Sudbury mineral range had few options other than seeking out "other" ores.

Iron ore was an obvious possibility because that mineral was a minor component of almost all local deposits. The somewhat richer ore of Hutton Township drew particularly close scrutiny after its discovery during the Vermilion River placer rush. Six or more companies were organized between 1903 and 1907; however, the isolation of the deposits precluded development until William Mackenzie and Donald Mann acquired a large block of Moose Mountain Mining Company shares in 1905. The noted railway magnates extended their Canadian Northern Ontario Railway north to Hutton Township, permitting on-site development to begin early in 1906.

The Moose Mountain mine produced 60,000 tons of ore by May 1908; unfortunately for the firm, an initial shipment was rejected by the prospective buyer because of excessive coarseness and low iron grades. Moose Mountain had little choice but to stockpile ore awaiting concentration, first by hand and then, after hydro-electricity was introduced in 1910, by magnetic cobbing. The improvements were insufficient to overcome weak iron markets so the mine was closed on 31 May 1911, throwing 250 men out of work. Many, however, found jobs installing a Grondal concentrating and briquetting plant at the site. Higher-grade iron resulted in the resumption of sales and, later, mining. The company's problems were far from over. Because the Grondal-process briquettes proved too costly to produce given the low price of iron, work ceased in 1914.[23]

In the face of this latest failure, Moose Mountain's still-hopeful backers financed yet more plant improvements. The prevailing optimism was especially evident in major expenditures on the adjoining company village. Once primitive Moose Mountain, renamed Sellwood in honour of a company founder, soon featured good streets and services, large public buildings, and various other amenities, all laid out on a simple but effective gridiron plan. Social aspects were not neglected: Moose Mountain officials provided administrative services and encouraged efforts by a variety of religious and secular organizations (but not, of course, the mine union movement). Social, administrative and financial commitments of this order seemed justified, for the war brought a stronger iron market. Regular operations were resumed in May 1916 and the pace of mining quickened through 1918. War's end, however, spelled disaster: fast-falling sales led to plant closure on 29 November 1920. Following the sale of stockpiled iron – bringing total sales to about 472,000 tons – the site was abandoned in 1923.[24]

Market forces also explained the rise and fall of a lesser-known mineral – feldspar. The scattered feldspar deposits south and east of Sudbury first attracted attention around 1918 because of increased demand by American pottery, glass and porcelain manufacturers. Work peaked in the mid-1920s, with pits in Dryden, Cleland and Dill Townships producing about 25,000 tons of raw feldspar. Small profit margins dictated lightly mechanized, summers-only quarrying; in 1926, for example, just over 100 men worked six pits. Unfortunately for the local industry, American buyers consolidated themselves in 1928 into two large consortiums, effectively ending the competition which had prompted expansion of the industry.[25]

While market forces naturally determined the fate of these more prosaic minerals, some optimistic – if not naive – mining speculators looked past such mundane considerations, confident that untold riches lay in the "next" deposit. No mineral, of course, offered greater hope than gold: as though to reinforce these dreams, one firm emerged with the capital, expertise and, most importantly, ore deposits required to achieve profitable operations. The Canadian Exploration Company acquired the just-discovered Long Lake gold property in 1908; the Drummonds of Montreal financed large-scale operations. Canadian Exploration established a road link to Naughton and a water-road link to Sudbury, then set about installing a cyanide-process plant and a combined shaft-pit mining operation. Good results quickened the pace: production by the 100-man crew doubled thanks to the addition of hydro-electricity and other improvements. The Long Lake mine thus became the highest-yielding gold mine in Ontario, an honour soon usurped by competition in the Porcupine. Worse, the small deposit was decimated by the pace of operations: 90 men were let go when the works closed in July 1916. Total production at Long Lake had exceeded 160,000 tons of ore, yielding 42, 265 ounces of gold and a little silver. Stripped of its valuable materials (apart from the tailings) the mine site was reclaimed by nature.[26]

However, the alluring rumours of gold usually proved more illusory. Little came of the intermittent search of the Lake Wanapitei pits, and a short-lived gold fever brought few benefits to the residents of Chelmsford. The Vermilion River gravels were never profitable despite the efforts of various entrepreneurs. Robert Ahn, for instance, returned to the Sudbury area in 1901, intent on working the river gravels in northern Hanmer Township. At "Dawson City" Ahn experimented with a "combined amalgamation and cyanidation process" mill, hoping to recover minute quantities of gold from dredged gravels. Large-scale work was undertaken from 1905 to 1908, but the Gold Gravel Dredging Company proved a failure. Interest in the placer beds persisted nonetheless. In 1909, the Vermilion River Gold Dredging Company began four years of fruitless endeavour; successive firms would find the gravel beds no less frustrating.[27]

The often-unreasoning clamour for gold had a counterpart in the continuing interest in the local "coal" (anthraxolite) deposits, which paid no heed to the failures of 1896-98. In 1910, the Chelmsford Coal, Gas & Oil Company undertook exploratory work on a 3,000-acre property; a shaft was driven some fifty feet before the investors came to their senses. More persistent efforts began in 1921 when the Balfour Development Syndicate shipped fifty-two tons of anthraxolite to Toronto, where it was marketed as fuel. Toronto mining speculators then took over, establishing at least five firms whose backers blithely ignored the refutations of the Ontario Department of Mines, the mining press and even young *Toronto Star* reporter Ernest Hemingway, which condemned the deposits as worthless. British-Colonial Coal Mines proved the most active firm, not least because of the promotions of "Professor" A.F.A. Coyne, who had earlier backed Winnipeg oil properties and who would go on to sell Montreal-area gas reserves. Coyne's efforts, and occasional shipments of samples, kept British Colonial in business until 1927, when the funds supporting its relatively extensive Balfour Township operations finally ran out.[28]

Ironically, but predictably perhaps, diamond drill cores taken during this renewed coal craze precipitated a larger boom, for they reawakened interest in the low-grade lead-zinc deposits located in the heart of the Sudbury Basin. Long known, and worked in the 1890s, these deposits were now more attractive because the booming manufacturing industries needed huge quantities of base metals. Prices naturally improved, with lead reaching a new peak in December 1924. These price gains, combined with improved reduction techniques that increased the worth of low-grade deposits, led the Treadwell-Yukon Company, controlled by the giant Bunker Hill & Sullivan Syndicate, to option 4,000 acres and commence tests in July 1925.[29]

Rumours of a "Billion Dollar Zinc Zone" soon inspired a frenzy of staking and company incorporation – more than forty new firms, worth over $100 million on paper, were organized. Many existing firms also launched exploratory operations; Dome Mines, for instance, optioned seven Hanmer Township properties. In all, more than seventy-five firms took some part in "the greatest concentrated diamond drilling campaign any district in Canada has ever seen." But the furore cooled by mid-1928 because most of the lead-zinc firms were "overnight promotions, inadequately financed, whose sole chance lay in immediately picking up a commercial ore body." This proved impossible, for the ores turned out to be very complex metallurgically and, far worse, of a lower-than-anticipated grade. The heavy overburden in the valley and boom-inflated land costs were additional difficulties; still, industry observers placed most of the blame on the investors themselves, who put forward money in response to wildly optimistic promotions and "abstract" geological theories about the lead-zinc zone. This ill-conceived speculation, the *Northern Miner* concluded, left "many of the hopeful Argonauts stranded. There are no doubt some fish on the beach as well."[30]

By early September 1928 the boom was over, with only Treadwell-Yukon, Sudbury Basin Mines and Towagmac Exploration's Lake Geneva Mining Company still actively seeking mineralized deposits. Sudbury Basin, however, soon turned its attention to local nickel deposits. Lake Geneva's considerable development in Hess Township – including a mine and a 100-ton concentrating mill – suspended production in 1931 pending improved metal prices.[31]

Treadwell-Yukon committed far greater funds and energy to its Errington mine; two shafts, a 250-ton pilot concentrating mill, a private utility line to carry electricity purchased from Wahnapitae Power, and other necessary mining facilities were in place by 1927. The once-sceptical *Sudbury Star* anticipated a "huge industry," especially after Treadwell-Yukon's capitalization was increased from $1.5 to $11.5 million in May 1927. The extra funds permitted the acquisition of an additional 2,000 acres of property, the bringing on stream of the concentrating mill on 21 April 1928 and, beginning in June, the sinking of a third mine shaft. The expanded operations boosted the workforce to about 400 men. Trouble at the mill slowed progress, but by year's end Treadwell-Yukon had announced major expansion plans, highlighted by a proposed base-metal refinery. The offer of a tax-free site for the refinery, unanimously approved by Sudbury town council, and the completion of a CPR spur to the Errington works were just two indications of the prevailing optimism.[32]

Fast-falling base-metal prices brought a changed attitude. In August 1929, Treadwell-Yukon announced a two-year delay in full-scale operations pending an assessment of the deposit's worth. Experimental mining and milling continued: through 1930 some 186, 170 tons of ore had produced about 12, 770 tons of zinc concentrate, 7, 840 tons of copper concentrate and 1, 940 tons of lead concentrate. Eight thousand tons of low-value iron pyrites concentrate and trace amounts of silver and gold added to the total. Unfortunately for the firm, zinc prices were very low because of African and Latin American competition, while copper and lead prices were falling. The local production (1928-31) was worth no more than $500,000. Poor income was compounded by the "disappointing" ore values uncovered during ongoing exploratory mining. As a result, Treadwell-Yukon determined that profitable operation of the Errington mine was unlikely and began curtailing production in December 1930. Its last local employees were released when the mine was abandoned on 15 December 1931 and allowed to flood. Treadwell-Yukon's failure dashed the last remaining hope for a significant mineral alternative to the nickel industry.[33]

SIGNIFICANCE OF THE "MINOR" MINERALS

Boom and bust, optimism and dismay, riches and ruin – what was the significance of the usually fruitless effort to exploit the minor minerals of the Sudbury area? Economic, physical and social evidence, though easily overlooked amidst the failure, fraud and greed, suggests that a maturing regional infrastructure owed much to the gold, lead and even coal seeker.

Economic gains were most apparent. Even when it only invested a portion of the "pickings" won from gullible investors, the mining industry brought real local progress. The labour-intensive nature of the mining pits, mines and plants meant work and wages outside the nickel industry. Sales of mining supplies, foodstuffs, timber and the like brought additional local income. Construction of roads, utility lines and mining camps created yet more prosperity; villages like Whitefish, Wahnapitae and Chelmsford received periodic boosts from the "lesser" mining booms. Sudburians, for their part, welcomed the economic blessings of these booms, while keeping an unfriendly eye on the more inflationary aspects of a fast-expanding economy. Wildly expansionary economies were of less concern to the mine brokers whose sales created the furore. Whatever the ethical merits of their promotions, the brokers drew attention and monies that would otherwise have bypassed the Sudbury area.

Individual economic gains derived from the lesser minerals were often fleeting; the physical consequences, in contrast, had a long-term impact. New transportation and communications corridors, new areas of settlement, new utility lines and other physical changes propelled the region towards a greater maturity, pushing back the wilderness that had prevailed just a few years earlier. New environmental damage – pits, deforestation, tailings – was no less apparent, although contemporary observers viewed the destruction as an acceptable price of progress.

For the most part the physical and economic impact of the "lesser" mining operations was a welcome boost to the local infrastructure. Social gains reinforced these changes: mining camps, no matter how short-lived, added to the human diversity of the region. Long-lived communities like Sellwood were especially important, serving as regional administrative and social service centres and as bases not only for order and control through church and lodge, but for blind pigs (unlicensed bars) and a host of other social alternatives. And even the most fleeting camp served as a temporary home for its crew, a social role hard to quantify but equally difficult to dismiss.

That there were profound differences between the social, economic and physical impacts of the various "lesser" mineral booms goes without saying. It is equally apparent, however, that the fascination with these minerals served to increase the maturity of the Sudbury-area infrastructure. Thus, individual enthusiasm and even gullibility remained influential in a setting supposedly dominated by giant nickel mining cartels. That was only fitting: Sam Ritchie's enthusiasms had "fathered" that giant industry; a similar confidence in the local resource base inspired a faith that it could happen again. That belief and the real gains that it produced suggest that the promoters and the prospectors – the Ahns, McCharles, Gordons, and Coynes – ought to be remembered not only for their failings but also for their enthusiasms. Likewise, the long-abandoned (or occasionally re-explored) pits – whatever the mineral – should be remembered not only as temples of failed greed but as invaluable stepping stones on the road to the modern Sudbury region.

Notes

[1] This paper is based upon portions of my dissertation "The Sudbury Area to the Great Depression: Regional Development on the Northern Resource Frontier" (Ph.D. dissertation, University of Western Ontario, 1988), which can be consulted for more detailed citations. This paper does not consider alternative mineral production dependent upon the nickel industry, most notably the extensive quartz quarries in Denison, Dill, Dryden, Snider and Waters Townships. Aggregate and clay pits are also omitted.

[2] The CPR-inspired interest helped to overcome doubts brought on by the spotty record of previous mining in Northern Ontario. Copper mining at Bruce Mines and the more spectacular income from Silver Islet notwithstanding, poor markets, isolation, technological failure and administrative shortcomings combined to prevent mining progress in Ontario's North. The same factors meant that the Sudbury deposits, though reported in 1856, were promptly forgotten. See Anna Wright, "The Canadian Frontier: 1840-1867" (PhD. dissertation, University of Toronto, 1943), 36-41, 76-77; Dianne Newell, *Technology on the Frontier: Mining in Old Ontario* (Vancouver: University of British Columbia Press, 1986), 59-69.

[3] The "Murray" location was obtained for $300, reinforced by threats of firing. William Montgomery interview, *Toronto Mail,* 25 November 1896; Francis C. Crean, letter to the editor, *Northern Miner, 2* October 1930; idem, "Sudbury Nickel Mines," *Montreal Gazette,* 3 December 1890; Aeneas McCharles, "The Nickel Range," *Toronto Mail,* 19 September 1891. On the earliest registered claims: Thomas Johnson, Toronto, to George Thompson, Renfrew, 18 February 1883, Crown Lands, "Letters Relating to Surveys," vol.15, reel 140, p.367. A succinct summary of early claims is W.J. Gorman, "Early Sudbury Discoveries and the Men Who Made Them," *Northern Miner,* 11 September 1930, 2-3.

[4] Aeneas McCharles was just one of a number of contemporary critics of the blanket claiming and rumoured improprieties by crown lands staff: *Globe,* 6 February 1886, 8 December 1890; *Toronto Mail,* 26 May, 3 December 1888, 3 December 1889; 2 April, 4 July, 6 August 1890, 1 May, 19 September 1891; *Toronto World,* 26 February 1897.

[5] The Crean-Harwood was later redeveloped as the Worthington mine. E. Odium, "The Mineral Wealth of Ontario," *Toronto Mail,* 26 June 1885; *Globe,* 20 September 1884; Northern Miner, 2 October 1930, 2; *Canadian Mining Review (CMR*) 3 (October 1885): 10; ibid., 3 (November 1885):

10; J.W. Evans, "The Early History of Mining in the Sudbury District," *Journal of the Canadian Mining Institute 7* (1904): 496.

[6] Evans, "The Early History," 496-97; Ontario, "Mines and Mining Operations," *Annual Report of the Bureau of Industries for the Province of Ontario* (1886), 378-79.

[7] Ritchie's speculative spirit and the history of Canadian Copper are well covered in Oscar W. Main, *The Canadian Nickel Industry: A Study in Market Control and Public Policy* (Toronto: University of Toronto Press, 1955). On the visit see the press reports, especially *Montreal Star,* 28 September 1886; *Ottawa Journal,* 27 September, 4 October 1886.

[8] The "yeasty and unsettled" boom was recalled in the *Toronto Star, 2* October 1902. Other reports include the *Globe,* 27 September 1886; *Perth Courier,* 5 November 1886; *Manitoulin Expositor,* 13 November 1886, 1 January 1887.

[9] The London (England)-based *Canadian Gazette,* 2 December 1886, was among the first publications to report the "tainted" ores. On the test results: Ontario, *Report of the Royal Ontario Nickel Commission* (Toronto: King's Printer, 1917), 38, 61-62.

[10] The Vermillion mine was also known as the Ranger, Tough, International and the Great Sudbury mine. The CMR and the Toronto, Ottawa and smalltown press carried dozens of

articles on the mine. One convenient source is Ontario, *Report of the Royal Commission on the Mineral Resources of Ontario* (Toronto: Warwick and Sons, 1890), 113, 309, 433-35. On Ranger: *Toronto World,* 8 March, 16 August 1888; *Monetary Times* 22 (21 September 1888): 329.

[11] *Toronto World,* 11 January 1888.

[12] *Ottawa Journal,* 18 October 1887. Additional reports on the boom include: ibid., 13 March; 6, 21, 24, 30 April; 26 May; 8, 27 June; 26 August; 2 October 1888.

[13] *Toronto Mail,* 22 September 1888.

[14] On the Simpsons: CMR 6 (August 1888): 93, 8 (July 1889): 86, 8 (August 1889): 101; *Monetary Times 22* (21 September 1888): 329; *Toronto Mail,* 22 September 1888, 11 June 1889. At least nine firms, worth over $500,000 on paper, were organized: Ontario, *Report of the Secretary and Registrar* (1888): 11-12; E. Kindle, "Gold Occurrences of Ontario East of Lake Superior," Canada Department of Mines, *Memoir,* no.192 (Ottawa: King's Printer, 1936), 92-99.

[15] Ontario, *Report of the Bureau of Mines* (*OBM*) (1891): 9, (1892): 236, (1893): 45-47, (1895): 261; *CMR* 12 (May 1893): 102, 12 (November 1893): 197, 13 (November 1894): 219, 14 (February 1895); *Canadian Mining Manual (CMM*) (1893): 351; *Globe,* 4 November

1893; P.E. Hopkins, "Ontario Gold Deposits," Ontario Department of Mines, *Report* (1921), Part II, 7.

[16] *OBM* (1891): 225; Geological Survey of Canada, *Report of the Mining and Mineral Statistics of Canada* (1890): 140s, (1891): 115s; *Canadian Mining and Mechanical Review (CM&MR)* 11 (February 1892): 7, 11 (October 1892): 169, 11 (November 1892): 184, 11 (December 1892): 200, 12 (January 1893): 3, 12 (February 1893): 15; *Toronto Mail, 22* August 1888, 11 June 1889, 2 October, 3 December 1890, 19 September 1891; *Ottawa Journal, 2* October 1889; *Sudbury Journal,* 26 March, 29 October, 5, 12 November 1891, 25 February, 21 April, 2 June 1892; *Globe,* 29 October 1892; Aeneas McCharles, *Bemocked of Destiny: The Actual Struggles and Experiences of a Canadian Pioneer, and the Recollections of a Lifetime* (Toronto: William Briggs, 1908).

[17] Besides reports in the OBM and GSC, *Mining and Mineral Statistics,* see: *Canadian Engineer* 1 (September 1893), 1 (December 1893), 2 (July 1894), 3 (September 1896), 4 (February 1897); *Sudbury Journal,* 18 April 1895, 29 October, 17 December 18%, 2 June 1898; *Toronto World,* 29 October, 17, 29 November, 1, 8, 15 December 1896, 16 January, 3, 23 February, 2 March, 30 April, 10, 26 May 1897; *Toronto Star,* 18 November 1896, 14, 19 De-

cember 1897; *Toronto News,* 19 January 1897; *Globe,* 21 November, 18 December 18%, 4, 20, 21 January, 23 March 1897; B.O. Dressier, "Geology of Lake Wanapitei Area," Ontario Geological Survey, *Report,* no.213 (1982): 103-12.

[18] *Globe,* 28 May 1898. Together, the Vermilion River placer and Lake Wanapitei booms saw more than a dozen firms organized, many with a "paper" value of $500,000 or more.

[19] Contemporary coverages of the placer beds include Arthur H. Gracey, "Placer Gold on Vermilion River," OBM (1897): 256-59; J.W. Evans, "The Gold Bearing Sands of the Vermillion [sic] River," *Journal of the Canadian Mining Institute 2* (1899): 105-7; A.P. Coleman, "The Vermilion River Placers," OBM (1901): 151-59.

[20] Examples of the reports include: for silver – *Ottawa Free Press,* 5 June 1888; *Toronto Mail, 2* October 1890; *Sudbury Journal,* 10 September 18%; for tin – *Toronto World,* 21 January 1889; for molybdenum – *Toronto World,* 11 January 1888; for cobalt – *CM&MR* 12 (February 1893): 20; for peat – *Montreal Gazette,* 18 June 1890; for diamonds – *Sudbury Journal,* 13 December 1900.

[21] *Toronto World,* 30 October 18%, provides the quote. For local praise see *Sudbury Journal,* 16 October, 5 November 1897. There is extensive press coverage of the find, beginning with

a first report in the *Toronto Star,* 10 July 18%. More reasoned coverage includes A.P. Coleman, "Anthracitic Carbon or Anthraxolite," Ontario Bureau of Mines, *Bulletin,* no.2, 18%; G.R Mickle, "Mineralogical Notes on Sudbury Anthracite," *Proceedings of the Canadian Institute* 1 (May 1897): 64-66. For operations and companies: OBM (1898): 141; CMM (1897): 36.

[22] Joseph Errington was among those who continued to test the area in the early years of the new century. See *Financial Post,* 18 February 1927, 20 January 1928. For details on the early work: *Monetary Times 22* (21 September 1888): 329; *Manitoulin Expositor,* 1 November 1890; *Toronto World,* 30 November, 29 December 1896; *Toronto Star,* 13 November 1896; *Globe,* 18 December 1896; *Sudbury Journal,* 21 April 1898, 6 February 1903; *OBM* (1899): 33; W.G. Uglow, "Lead and Zinc Deposits in Ontario and in Eastern Canada," *OBM* (1916): Part II, 17.

[23] Other firms included National Iron (1903); Hutton Mining (1903), Vermilion River Iron (1906); North Range Nickel & Iron (1906); Onaping Iron (1907). On Moose Mountain: OBM (1907): 16, (1908): 22, (1909): 90, (1910): 87, (1912): 112, (1913): 28, 106, (1914): 112, 123; F.A. Jordan, "Moose Mountain Iron Mine," *Canadian Mining Journal* 33 (1 December

1912): 807-10; E. Lindeman, "Moose Mountain Iron-Bearing District, Ontario," Canada, Mines Branch, *Report,* no.303. There are numerous newspaper reports.

[24] The annual volumes of the Ontario Bureau/Department of Mines provides much detail. Also: *Sudbury Star,* 20 December 1916, 23 January 1918, 18 April, 11 August, 7 November 1923, 19 March 1924, 16 February 1927, 23 March 1932; E.L. Bruce, "Future of the Ontario Iron Deposits," *Queen's Quarterly* 23, no.2 (1924): 177; Colin Clay, "Capreol and Surrounding Area," *Historic Sudbury* 1 (1979): 44-46.

[25] OBM (1919): 49, (1920): 38, (1923): Part VI, 27, (1924): Part VII, 21, (1925): 7879, (1926): 82-83, (1928): 73; *Sudbury Star,* 15 November 1922, 2 May, 17, 28 November 1923, 22 November 1924, 4 March 1925, 31 December 1926; *Northern Miner,* 11 April 1925, 17 June 1927; Hugh S. Spence, "Feldspar," Canada, Department of Mines, Mines Branch, *Report,* no.731 (1932): 57-60, 92-93.

[26] OBM (1909): 45, 85, (1910): 83, (1911): 9, (1912): 112, (1913): 101, (1914): 112, 122, (1915): 103, (1916): 78, (1917): 10, 89; M.B. Baker, "Long Lake Gold Mine, Sudbury District," OBM (1917): 258-74; E.L. Bruce, "Precious Metal Mining in Northern Ontario," *Journal of the Canadian Bankers' Association* 33 (April 1925): 354; J.H. Stovel, *A Mining Trail 1902-*

1945 (1956). Stovel managed the mine for several years and discusses it in this autobiographical account.

[27] See the general reports by Kindle and Hopkins cited above. Also: J.B. Gordon et al., "Gold Deposits of Ontario," Ontario Geological Survey, *Mineral Circular,* no.18, Part II. The placer beds yielded "just enough gold to tantalize but not sufficient to repay the hardest worker," *Saturday Night,* 23 May 1908, 5. General coverages include V.K. Prest, "The Pleistocene Geology of the Vermilion River System Near Capreol, District of Sudbury, Ontario (with special reference to placer gold)," Ontario Department of Mines, *Preliminary Report,* no.2 (1942): 1; S.A. Ferguson and E.B. Freeman, "Ontario Occurrences of Float, Placer Gold, and Other Heavy Metals," Ontario Geological Survey, *Mineral Circular,* no.17 (1978): 178-81.

[28] On the 1911 effort: OBM (1911): 50; *Sault Star,* 1 December 1910; *Sudbury Star,* 8 March 1911. The later search is reported by the Department of Mines. See also *Sudbury Star, 2* April 1921, 27 September 1924, 4 March 1925, 19 May 1926; *Financial Post,* 11 June 1921, 22 September 1922, 2 April 1926; *Saturday Night, 22* April, 1, 22 July, 5, 19 August, 23, 30 September, 13 October 1923, 9 February 1924, 11 July 1925; *Northern Miner,* 5 May,

4 August 1923, 20 February, 31 July 1926; Ernest Hemingway, "Search for Sudbury Coal a Gamble," *Toronto Star,* 25 September 1923.

[29] For details on the lead-zinc zone: A.G. Burrows and H.C. Rickaby, "Sudbury Nickel Field Restudied," ODM (1934): Part II. Work through 1925 is detailed in ODM (1923): 27, (1926): 30, (1927): 59; *Financial Post,* 31 July, 14 August 1925; *Northern Miner,* 15 August 1925; *Saturday Night,* 11 June 1927.

[30] Nearly every issue of the *Financial Post, Northern Miner* and *Sudbury Star* circa 1927-28 carried news or advertisements concerning the lead-zinc zone. Information on the firms is available in the annual list of incorporations published by the Department of Mines, as well as Canada, "Synopsis of Letters Patent," *Financial Post Survey of Mines,* federal and provincial *Gazettes,* and in advertisements. For overviews of the excitement: *Commercial and Financial Review* (1926): 16; *Canadian Annual Review* (1926-27): 328; *Northern Miner,* 6 September 1928; *Financial Post,* 15 June 1928, 19 April 1929. The quotes, in order, are from *Toronto Telegram,* 25 December 1926; *Financial Post Survey of Mines* (1929): 143; *Financial Post, 7* December 1928; *Northern Miner,* 6 December 1928.

[31] The fortunes of Sudbury Basin and Towagmac are most conveniently traced through the De-

partment of Mines *Reports* and the annual *Financial Post Survey of Mines.*

[32] ODM (1927): 30, 144, (1928): 28, 159, (1929): 27, 83-84, (1930): 26, 79-80; *Sudbury Star,* 3 March 1926, 23 February, 4 June, 24 August 1927, 16, 20 June, 8 September 1928, 31 August 1929; *Northern Miner,* 24 March, 21, 28 April, 13, 20 October, 17 November 1927, 12 January 1928; *Saturday Night,* 28 May 1927, 23 June 1928, 12 January 1929; *Financial Post,* 28 May 1926, 29 April, 6, 27 May, 17 June, 1, 29 July, 18 November 1927, 6, 20 January, 17 February, 16 March, 27 April, 22 June, 9, 16 November 1928, 11 January, 22 August, 12 September, 19 December 1929.

[33] ODM (1931): 64-65, (1932): 64; *Financial Post,* 22 August, 10 October, 19 December 1929, 20 March, 24 April, 20 November, 25 December 1920, 26 March, 19 November, 26 December 1931; *Northern Miner,* 13 March, 6 November, 25 December 1930. Ore and metal production is calculated from Department of Mines and press reports.

Chapter 4

The Image and the Reality of Life in a Northern Ontario Company-Owned Town

Eileen Goltz

The company town, which has long been an integral part of Canadian history, has become an anachronism. As early as the late 1950s companies began to transfer company-owned land, housing and amenities to private and public ownership, and to replace company supervision of town activities with freely elected municipal councils. While they existed, company towns were the subject of images and realities that did not always coincide. The most pervasive reality of life in any company town was company control, which, whether it was harsh or mild, obvious or subtle, was always present, and established parameters for the realities of life within the town. The images, especially those perceived by people living outside the company-town environment, were fodder for the genesis of myths that often portrayed the company town and life therein in a manner that differed from reality.

This paper explores five of the many images associated with the company-town phenomenon. The financial aspect of company-owned housing, the company's response to unemployment in the town, the attitude of residents towards the company town, and the relationship between status and address will be explored briefly, while the cohesiveness of the company town will be discussed in greater detail. These images will be discussed within the context of the former company town of Copper Cliff, near Sudbury, and it will be suggested that only one of the images bears any relationship to reality.

Copper Cliff was a company town from 1886 until 1973, when it was annexed by the City of Sudbury. The classic rationale for construction of a company-owned town applied to Copper Cliff. It was inaccessible and remote from large centres of population. Over its eighty-seven years as a company town, Copper Cliff served as a mining, smelting and refining centre, and as the administrative headquarters for the owning company – first the Canadian Copper Company (CCC) and later the International Nickel Company (Inco). During this period, the company provided housing and the urban amenities required to retain those elements of the workforce it considered important. These were English-speaking technical, professional, managerial and skilled personnel who were primarily English-speaking Canadians or immigrants from the United States or Great Britain.

The company catered to these Anglo-Saxons by housing them in a separate area or enclave known as the Copper Cliff settlement, which became the *de facto* company town. Within this settlement the company provided living accommodation and recreational, medical and shopping facilities. The company did not provide housing or other amenities for non-Anglo-Saxons. They were expected to build their own homes on land leased from the company, in settlements surrounding the Copper Cliff settlement. Although non-Anglo-Saxons were permitted to use the medical and shopping facilities within the Copper Cliff settlement, they were not welcome in the company-owned housing, nor in the recreational buildings, constructed and sponsored by the company. Among the latter were the Gorringe Club, built in 1900, and its replacement, the Copper Cliff Club, built in 1916. Both clubs offered facilities that were not available elsewhere in town.[1]

One of the most obvious manifestations of control in any company town was the spatial segregation of housing according to ethnicity and occupation, a practice in which the rigid corporate hierarchy is clearly reflected in town planning and is imposed on town residents. Thus, the commonly held belief that the status of a person living in a company town was indicated by his or her address was a reality.[2] This image is accurate and applied in Copper Cliff where the CCC introduced ethnic segregation by separating the Anglo-Saxon residential area from the Finnish and French-Canadian sections. The three ethnic groups had all lived in shacks and shanties in Shantytown, south of the Copper Cliff Mine, until the fall of 1886, when the CCC moved the Anglo-Saxon workers into company-built houses in the newly surveyed Copper Cliff settlement, north of the mine.[3] This segregated Anglo-Saxon from non-Anglo-Saxon employees, and delineated an area within which the company could cater to an elite. The Copper Cliff settlement became, and remained, the true company town, wherein the company provided recreational and medical facilities, and encouraged the growth of a business and retail section. Residence in the new settlement was restricted to This map (4.1) was redrawn from Map Nos.824 and 825, originally prepared by Alfred E. Barlow, and appearing in the Geographical Survey of Canada Report, 1904, Part H. A number of small settlements made up the company town of Copper Cliff in 1904. The oldest of these were Shantytown and the Copper Cliff settlement, both of which had grown up around the Copper Cliff Mine. The Evans Mine settlement was almost as old, but by 1904, due to mine closure, was disappearing. The Johnson Extension and the East Smelter settlement had begun to deteriorate following the closure of the East Smelter and the relocation of smelting activities to the West Smelter, near the Crow's Nest settlement. This latter settlement was being displaced by the expansion of Little Italy. The Orford settlement was gradually being abandoned as workers moved into more central locations.

Map 4.1 Copper Cliff in 1904

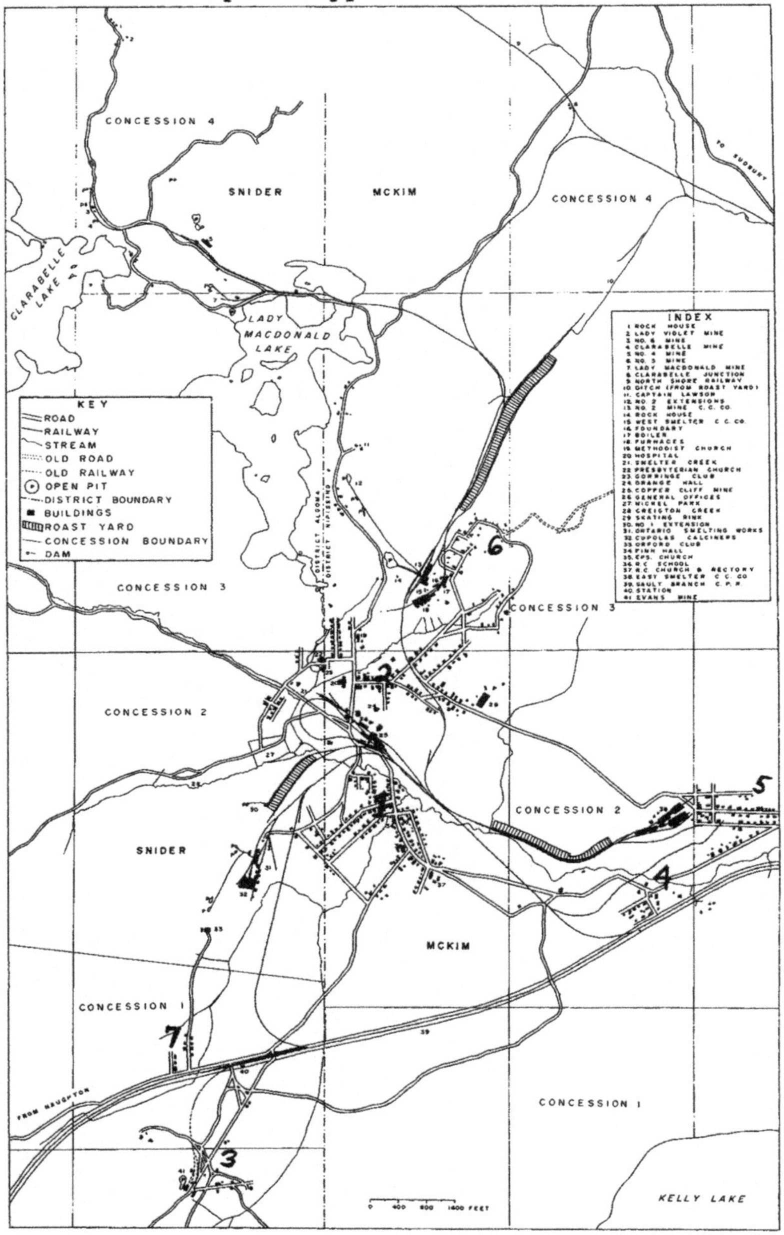

The settlements as they appear on Map 4.1 are explained in the legend below:

1. Shantytown
2. The Copper Cliff settlement
3. The Evans Mine settlement
4. The East Smelter settlement
5. The Johnson Extension
6. The Crow's Nest, and the expanding Little Italy
7. The Orford settlement

Anglo-Saxons, who rented company-owned houses, leased land on which they built their own homes, or constructed commercial buildings.[4]

The CCC made no effort to segregate the remaining non-Anglo-Saxon groups from one another, and readily leased lots for private building purposes to the Finns and French Canadians who continued to live in Shantytown and to the Poles who joined them later. The names of some streets in Shantytown reflect the ethnicity of the people living there. Two of the most obvious are Finland and Poland streets. The same segregationist philosophy prevailed when other sections of the Copper Cliff area were settled. Anglo-Saxons lived in proximity to one another, usually in company-owned houses, while non-Anglo-Saxons settled on nearby land leased from the company, in what were often ethnically mixed neighbourhoods. House rentals, land leases, and employment provided the company with powerful weapons of social control over the workforce. The failure of tenants or lessors to comply with company wishes could, and occasionally did, result in dismissal from the job and eviction.

By 1900 the CCC had established a form for the social landscape of Copper Cliff. The Copper Cliff settlement was central with Shantytown, populated by Finns, Poles and French Canadians, to the south; the East Smelter settlement with its small contingent of Anglo-Saxons and French Canadians, at the eastern extremity of company-owned land; and Little Italy, developing as the home of the Italian immigrant population and a few French Canadians, to the north. The street names in Little Italy (Lombardy, Milan, Pietro) reflected the background of the dominant group in the settlement. Directly east of company jurisdiction was the Johnson Extension, the privately owned fringe development with a population of Ukrainians, other East European immigrants and a few French Canadians. The incorporation of the Town of Copper Cliff in 1901 united these diverse settlements under one elected council. Elective municipal government, however, did not eliminate company control; elections were few, and only occurred in the early years of incorporation. A company-approved slate of candidates was usually acclaimed. At this early stage, address only indicated whether or not a person was Anglo-Saxon. Although company executives had begun to cluster on streets west of the business section, the CCC had not consciously introduced occupational segregation.[5]

Map 4.2 Copper Cliff in 1920

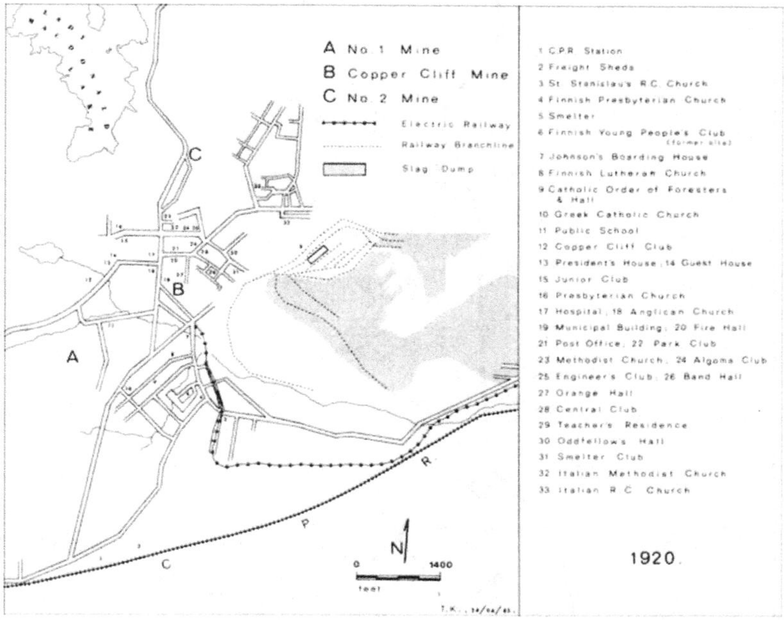

By 1920 the Evans Mine and Orford settlements had disappeared, the Crows Nest had expanded eastward to become Little Italy, and few houses remained in the East Smelter settlement or the Johnson Extension. The Copper Cliff settlement it (the de facto company town), which was expanding southward along Power Street, was bounded on the north by Little Italy and on the south by Shanty town. The encroachment of the slag-dump on the Old East Smelter settlement is indicated on the map below.

Source: Tom Kujanpaa prepared this map, using several maps held by the Copper Cliff Museum.

Map 4.3 Copper Cliff in 1972

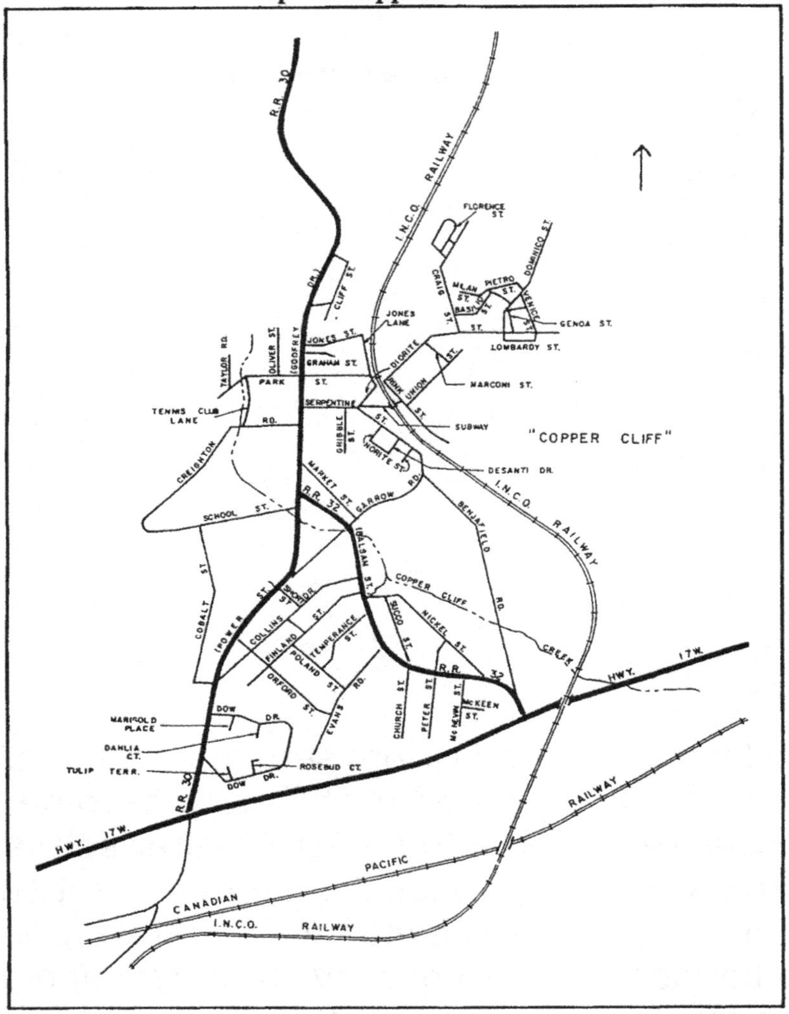

This map indicates the street layout of Copper Cliff at the time of annexation. Except for the addition of new streets, the street plan was altered very little over the years from 1920 to 1972.

Source: Eileen Goltz, "Copper Cliff: The Pioneer Period" in F.A. Peake, ed., Industrial Communities of the Sudbury Basin: Copper Cliff, *Victoria Mines, Mond, and Coniston* (Sudbury, 1986), 21.

The policy of occupational segregation came with the 1912-14 house-building program which put residents with similar occupations in houses of a specific size and quality on particular streets in the Copper Cliff settlement. This succeeded in dividing the Anglo-Saxon residential districts into occupational clusters. The company thus made visible a hierarchy based on occupation, and imposed social divisions on the community, although the social distance from one group to another was often far greater than the physical distance.[6] Address indicated status, and status was directly relative to ethnic background and to position within the company's occupational hierarchy. Although the situation within non-company towns is similar, it is imposed by mutually accepted constraints, not by an owning company. In an independent town, people are normally free to move from one section to another. This freedom is not available in company-owned towns, where companies can and do dictate where individuals may live. In Copper Cliff, movement within the town was restricted both by the availability of accommodation and by company policies which established parameters for access to accommodation.[7]

Thus, the image and reality concerning address and status coincided in Copper Cliff, and the same reality was present in other Canadian company towns. Occupational and ethnic segregation were integral to the company-town phenomenon, and when these towns were holistically planned, as was Temiskaming, segregation was built in as part of the planning.[8]

The image of a company town where the company provides employees with housing at a financial loss to itself is a myth created by the companies who strived to portray themselves in a positive light, and perpetuated by some Canadian researchers of the company-town phenomenon who have often been misled by the low rents which prevailed in company towns relative to those charged for similar, privately owned accommodation in nearby towns. One of the best-known writers on the topic is H.W. Walker, one of the authors of *Single Enterprise Communities in Canada,* published in 1955 by the Institute of Local Government, Queen's University.[9] This publication remains one of the most important comprehensive studies of the Canadian company-town phenomenon.

While the understanding that companies gained a financial advantage through house rentals is not readily accepted in Canada, it forms one of the hallmarks of the study of company towns in Britain, and is accepted in the United States.[10] The expectation of profit is a natural consequence of any business venture. Companies are not normally philanthropic organizations, and those that owned Copper Cliff were not exceptions. Like most companies, they were answerable to directors and shareholders who would not have tolerated fiscal irresponsibility. Both the CCC and Inco perceived company-owned housing to be a means of maximizing profits, and therefore viewed the operation of a company town as an investment from which a reasonable profit was expected. There is ample evidence in company correspondence of the company's intention to realize a profit from its house rentals.

In 1888 the CCC built log houses at a cost of $100 each, and rented each of them for $5 per month, or for $60 per annum.[11] This was a high rent for houses in Copper Cliff which during that period usually rented for $1 to $3 per month, and ensured that the capital costs of building would be recovered in less than two years. Thereafter, the rental income would have been pure profit, especially since no regular maintenance program existed, then or later. In 1889 Dr. E.D. Peters, Jr., the general manager of the CCC at Copper Cliff, explained to his superiors in Cleveland that the company should expect to receive 20 percent interest for the small houses then being constructed.[12] In 1903 A.P. Turner, the CCC president who lived in Copper Cliff, pointed out that the houses he wished to build could be rented for enough "to pay a very good interest on the investment."[13] In 1913, during a house-building program which instituted occupational segregation, the assistant to the president noted that rentals for the small houses being built for semi-skilled employees would return 11 percent per annum on the original investment.[14]

The expected return on the housing investment in Copper Cliff was not uniform at this time. Those employees renting less prestigious houses contributed disproportionately more to the company's profit than did those occupying better houses. Before the 1912-14 building program had been completed, the CCC's head office demanded a fixed return of at least 6 percent on all housing. The order caused some dismay among officials in Copper Cliff who had already rented accommodation to some skilled workers at rates which did not return that much.[15] The varying rents charged for the houses Inco built from 1947 to 1949 indicate that this policy was never rescinded. The escalated building costs experienced in 1948 and 1949 were reflected in the higher rents charged for the houses built during those years compared to the rents charged for similar houses built in 1947.[16] There can be little doubt that house-renting, contrary to popular belief, was profitable for the company. This was neither a positive nor a negative practice. Rather, it indicated the pragmatic nature of the company whose prime objective had to be the maximization of profits.[17]

The image that company towns were prosperous and that companies cared for their tenants during times of economic recession does not necessarily contain contradictory concepts. The myth is based on the reality that few inhabitants of company towns ever became welfare recipients. Another reality, however, is that companies, including those which owned Copper Cliff, did not normally allow the unemployed to remain in company housing. Anyone who was not among the employed was expected to move out of the company town. Companies, thus, were able to use their housing policies to reduce the amount of money expended for welfare assistance by shunting potential welfare recipients to other jurisdictions. One of the most obvious examples of this policy occurred in Copper Cliff when a large number of company-owned houses remained vacant during the shutdown of Inco operations which resulted from the economic depression following World War I.[18] Since unemployment was not allowed to exist in the company town, the perception of prosperity was encouraged, and since few inhabitants ever collected welfare benefits, the perception of a company willing to care for its employees was also encouraged.

In a company town, the category "unemployed" included such potential recipients of welfare assistance as widows and retirees. In Copper Cliff it was rare for either widows, retirees or the unemployed to remain in company housing. To be sure, there were a few exceptions who lived with a family member who was employed. In some instances the company would permit transfer of the house lease to the employed family member. This was not, however, a usual practice, and the company did not normally permit unemployed people to move in with other families.

Only those people who leased company-owned land and owned their own homes were able to remain in Copper Cliff following retirement, unemployment or widowhood. These people lived in the non-Anglo-Saxon enclaves surrounding the *de facto* company town, and were ignored by company officials.[19]

In company-town literature, life in the town is depicted as paternalistic and unacceptable to company employees.[20] Those who have studied the phenomenon in the United States have been most vocal in their denunciation of company towns. They decry the excessive control wielded by companies over the lives of townspeople, and express concern that company towns exist beyond the pale of the democratic process, thus rendering their inhabitants powerless.[21] This negative perception has affected the establishment of company towns in Canada. For example, E.C. Drury, who was the premier of Ontario from 1919 to 1923, was so firmly opposed to the concept of company towns that he insisted the pulp and paper town of Kapuskasing, in northeastern Ontario, be established as an independent community.[22]

Most Canadian writers have been more positive concerning company towns than their American counterparts have been. They regard the establishment of such communities as a necessity, and studies of individual Canadian company towns often state that the people living in those towns accepted company paternalism and considered it to be evidence of a caring company. These inhabitants perceived company housing, municipal services and other amenities to be beneficial, and they were content to overlook what to others might be interference in daily activities.[23]

Such was the case in Copper Cliff. Possibly the best evidence that this was so is the fact that the companies responsible for Copper Cliff never had enough housing available to satisfy the demand. There was always a backlog of requests for company houses, and those who lived in them did not normally vacate them willingly.[24] If the tenants chafed under company control, they did so quietly for there is little evidence of discontent. The low rents and the amenities provided by the company made company control tolerable in whatever guise it appeared. Life in the company town of Copper Cliff was very acceptable to the company's tenants. A large file of applications from employees seeking to rent company-owned housing is kept in the Inco Archives and provides evidence of the popularity of company-town living. Applicants were asked to indicate their reasons for wishing to live in Copper Cliff, and while the modest rents charged for company housing were important, other factors also assumed an importance. Many applicants wanted access to the church, school and medical facilities they believed were superior to the ones they were experiencing in Sudbury. Others perceived that in Copper Cliff they would enjoy an urban environment where the company provided, and was responsible for, everything. The applicants had a vision of paternalism as protection, and of company-town living as idyllic. There were not, however, sufficient houses 6 available in Copper Cliff to satisfy the demand, which increased yearly, and company officials were able to choose tenants from among many applicants. There was an indication among the applications that, while people wanted their requests granted immediately, they were willing to wait until a house became vacant. There were no suggestions that the company should construct more housing.[25]

The image of company towns as cohesive units is a myth based on the perception that company towns were insular entities, despite the metropolitan thesis and its emphasis on the connection between dominant urban centres and smaller communities in the hinterland.[26] The world outside the company-town environment tends to regard these small communities, partly because of their size, as cohesive structures. Since the company speaks for the town, the impression of cohesion is enhanced by the lack of dissenting opinion. Town spokespersons are company designates who, perceiving the town to be part of the company, present the town as an encapsulated community, bound to the company rather than to a metropolitan centre elsewhere.

In Copper Cliff, each of the main ethnic and language groups – the Anglo-Saxon, Finnish, French Canadian, Italian, Polish and Ukrainian – lived separately. Although each appeared cohesive and self-contained during its early stages, none remained so for long. Cohesion among the several groups living in Copper Cliff was non-existent. Each group interacted with similar ethnic or language groups outside Copper Cliff, rather than with dissimilar groups within the company town. To explore the myth of the cohesive company town, it is necessary to consider separately each of the six main population groups noted above. This examination will indicate that even as the town aged, the segregation begun in 1886 continued, and that cohesion, absent among the population groups in 1886, continued to be absent.

Anglo-saxons

The first group to be considered is the dominant Anglo-Saxon community. Anglo-Saxons had, from the very beginning of settlement in the area, comprised the largest proportion of Copper Cliff's population.[27] The figures published in the 1901 census indicate that there were 1, 633 Anglo-Saxons among a total population of 3, 699. Thus, Anglo-Saxons formed approximately 44 percent of the population of Copper Cliff in 1901.[29] The census years 1911-71, the last year Copper Cliff was enumerated as a distinct place, provide a more accurate assessment of the proportion of Anglo-Saxons in the population, and that proportion is illustrated above in Table 4.1.

Table 4.1 Number and Percentage of Anglo-Saxons Living in Copper Cliff[28]

Year	Total	Anglo-Saxon	Percent of Total
1911	3,082	1, 168	37.89

Year	Total	Anglo-Saxon	Percent of Total
1921	2, 597	1, 402	57.98
1931	3, 173	1, 370	43.17
1941	3, 732	1, 962	52.57
1951	3, 974	2, 170	54.63
1961	3, 600	1, 839	51.08
1971	4, 100	2, 225	54.27

The Anglo-Saxons formed the dominant group socially, economically and numerically, and their segregation from others in the village was emphasized by the company policy which clustered within their Copper Cliff settlement the services and the institutions they used. The business section, with its retail stores, boarding houses, bank, hospital and post office, was located within the enclave, and both the public and secondary schools were built in the Copper Cliff settlement.[30] The churches attended by Anglo-Saxons – the Presbyterian, Methodist, Anglican and (later) the United – were clustered there also, while the Roman Catholic church was located in Shantytown.[31] Although some Anglo-Saxons were Roman Catholic, the officials who allocated lots for leasing purposes were not. They associated Roman Catholicism with the non-Anglo-Saxon portion of the population and, hence, located the church in the non-Anglo-Saxon part of the town. Since the company owned all land within its town, it could choose to whom it would lease lots, and which lots it would lease. Since the company reviewed all building plans prior to construction, it was well able to impose its philosophies on parties wishing to erect any structure. The religious divisions within the population of Copper Cliff indicated a lack of cohesiveness similar to that which existed in most non-company-owned communities.

The segregation and elitism of Anglo-Saxons were further emphasized by company policies which provided recreational facilities for the exclusive use of their community. In 1900 the company built the Gorringe recreational club house, and in 1916 replaced it with the Copper Cliff Club. Non-Anglo-Saxons could not obtain membership in these clubs, nor were they permitted access to the facilities.[32] The Anglo-Saxon population did not depend solely on company-sponsored recreation, however, and the plethora of Anglo-Saxon fraternal, benevolent and social organizations which flourished in Copper Cliff provided members with ties to similar associations in other municipalities and exemplified the diversity which existed among those living in the Copper Cliff settlement. None of these organizations welcomed non-Anglo-Saxon members.

Anglo-Saxons in the Copper Cliff settlement had originally held positions at all occupational levels. However, with the house-building program of 1912-14, the company had stopped providing houses for labourers and semi-skilled workers.

The exclusion from company-owned housing of these groups indicated the initiation of a movement to restrict residence in the settlement to supervisory, skilled and administrative-level employees, and to any other people whom the company deemed important at different times, such as medical personnel, teachers and merchants.

The economic recession which followed World War I resulted in decreased mineral production, reduced workforces, a shut-down of all Inco operations in the Sudbury District, and vacancies in many company-owned houses. Since company policy did not allow people who were unemployed to remain in company-owned housing, there was a dispersal of the Anglo-Saxon population.[33] The introduction in 1915 of a street railway system, which provided an inexpensive form of transportation between Sudbury and Copper Cliff, meant it was no longer necessary for CCC employees to live in Copper Cliff. This reduced the need for the company to provide housing.[34] The semi-skilled and labouring elements of the Anglo-Saxon workforce were thereafter expected to live in Sudbury and to commute daily to work in Copper Cliff via the street railway.

The formation of the United Church of Canada in 1925 provided another example of the lack of cohesion that existed within the Anglo-Saxon community. Some of the Presbyterians, rather than attend the newly organized church in Copper Cliff, joined forces with dissident Presbyterians in Sudbury to build a new church.[35] Religious differences were sufficiently important to the Copper Cliff Presbyterians that they allied themselves with a Sudbury group rather than with people from within the enclave.

In 1929 Inco and the Mond Nickel Company (Mond), the other large mining corporation operating in the Sudbury District, merged. Following the merger, Inco moved the Mond supervisory personnel to Copper Cliff, thereby increasing the proportion of administrative people living in the town. It then began to remove the houses it considered inappropriate for the occupational levels it was willing to accommodate.[36] As production increased in 1935, Inco expanded the Anglo-Saxon contingent of its workforce. Aware that Sudbury was only a streetcar ride away, the company decided to provide amenities in Copper Cliff superior to those available in Sudbury, a centre which had, since 1930, enjoyed city status. Among the new amenities were a community hall and an artificial ice arena – facilities that Sudbury was not able to provide. The new recreational buildings were open to all Copper Cliff residents, even non-Anglo-Saxons, and to company employees living in Sudbury. Their lack of exclusivity rendered the Copper Cliff Club, by comparison, even more exclusive.

A house-building program from 1947 to 1949 continued the practice of providing company-owned houses only to supervisory, skilled and administrative employees and other people the company believed should live in Copper Cliff. This was the last building program undertaken by the company in the town, and at its conclusion some non-Anglo-Saxons were being accepted as tenants. This was pragmatism, not tolerance. If the skills or education of the person concerned were required by the company, ethnicity was overlooked. Such had not always been the case. At the turn of the century experienced Finnish and Polish miners, although in demand as employees, lived in Shantytown, while the company provided accommodation in the Copper Cliff settlement for Anglo-Saxon miners. This practice continued until the late 1940s. After 1950, the Anglo-Saxon enclave in Copper Cliff stagnated. No more house-building occurred, and few houses were vacated by the tenants. Thus there was no renewal provided by the movement of new residents into town. The company had, by the early 1960s, lost interest in retaining Copper Cliff as a company town and did not object when in 1968 Sudbury petitioned the Ontario Municipal Board to annex the community.

Finns

The experience of the non-Anglo-Saxon groups was different, for unlike the Anglo-Saxon group, which was never cohesive, the others began with an internal cohesion that eroded over time. In the case of the Finns, statistics concerning the population are elusive because Canadian census enumerators used a political-geographic definition to indicate ethnicity. Thus, since Finland was a province of Russia from the early 1800s until 1918, emigrants from there were counted as Russians until the 1921 census, when the designation Finnish was used for the first time. The term "Finnish" continued to be used by Canadian census officials until the 1951 census when it was eliminated, and Finns were enumerated as "others." There were, in 1901, approximately 401 Finns (Russians) living in Copper Cliff, and they formed 10 percent of the total population.[38]

Table 4.2 Number and Percentage of Finns Living in Copper Cliff[37]

Year	Total	Finnish	Percent of Total
1911	3,082	537*	17.42
1921	2, 597	267**	10.28
1931	3, 173	491	15.47
1941	3, 732	394	10.55

* Russians. ** There were four Russians living in Copper Cliff in 1921.

Finnish immigrants, having arrived in the Copper Cliff area from Finland via the United States in 1885, had been among the original employees hired by the CCC in the spring of 1886. They remained in Shanty-town after the company removed the Anglo-Saxons, and by 1895 their community was sufficiently cohesive to form a local branch of the Finnish Temperance Brotherhood of America, to build a hall in Shantytown, and in 1897 to establish a Finnish Lutheran Congregation. The local temperance society had ties through its parent body with similar groups in Finland and in the United States, and the itinerant American clergy who held services in Copper Cliff provided a conduit to Lutherans living elsewhere.[39] In 1908 the Finnish

Lutheran congregation built a church which was partially financed from the proceeds obtained through the sale of the Temperance Hall.

By that time, however, it was evident that the community had lost much of its earlier cohesion. The formation of the Young People's Society in 1903, an organization which attracted a large following, indicated the increasing diversity of the community. The new society built a hall in Shantytown and, unlike the Temperance Society, entertained local audiences – both Finnish and non-Finnish – and held fund-raising activities. In 1912 the Young People's Society joined the Socialist Party of Canada, causing a rift within society ranks and within the Copper Cliff Finnish community. The two factions – the socialists and the non-socialists – never reconciled. The larger socialist contingent of the Young People's Society and the smaller dissenting group within the society engaged in a series of lawsuits to gain control of the hall and of the society's treasury. During the litigation neither side was able to use the hall, and in March of 1915, when the Ontario Supreme Court decided in favour of the socialist members, the building and the money were transferred to them. Four days later the hall, which was fully insured, burned amid claims of arson which prompted an inconclusive police investigation. The CCC, anxious to remove the socialist influence from its town, refused to lease land to the Young People's Society, and began to ease known socialists out of the town.

The society then purchased a lot in the Johnson Extension, the privately owned fringe development, and erected a new Finnish Hall which served as a place of entertainment for both Sudbury and Copper Cliff.[40]

Cohesion was further eroded in 1913 when a Finnish Presbyterian Mission leased a building from the company and, with a Finnish-speaking clergyman and parish workers, began to provide religious, educational and social services to the Finnish population. The mission, financially assisted by the company, promoted assimilation through the teaching of English and the promotion of Canadian traditions, and in the process alienated the Presbyterian Finns from other Finns in the community.[41] The company's support of the mission was pragmatic and arose from a keen awareness of the problems that could occur when groups of non-English-speaking people working in a dangerous environment did not comprehend instructions.[42]

The post-World War I economic recession was responsible for a movement of Finns into Sudbury where they formed the basis of a Finnish community. The Finnish population in Copper Cliff was already fragmented when, in 1938, the former Finnish Presbyterian Mission, by then the Finnish United Church, was closed.[43] Members of the small congregation were dispersed between the Copper Cliff United Church and All People's United Church in Sudbury, a church established to minister to Sudbury-area immigrants. The deeply divided Finnish community in Copper Cliff stagnated.

French Canadians

The 1901 Canadian census indicates that there were approximately 1,007 French Canadians living in Copper Cliff, and that they formed 27 percent of the population.[44] Both the number and proportion of French Canadians decreased in 1911, and as the population of Copper Cliff expanded slowly from 1921 through 1951, so also did its French-Canadian component. The increased numbers of French Canadians, however, resulted from natural increase, not migration.[45] In 1961 the total population of Copper Cliff decreased, while the French-Canadian element increased slightly, producing a small percentage gain. The population expansion noted in the 1971 census reflected the construction, by a development company, of rental units at the southern edge of the town. Most of the townhouses and apartment suites built in that area were leased by Anglo-Saxons and French Canadians.[46]

French Canadians, along with Finns, had been members of the earliest work crews in the Copper Cliff area, and many remained following incorporation. Most of these French Canadians had originated from Quebec and were closely allied to the French-Canadian community in Sudbury. In 1889 they and other Roman Catholics in Copper Cliff, led and counselled by Sudbury-based French-speaking Jesuit priests and supported by the French-Canadian community in Sudbury, demanded that the school, then being built by the company in Copper Cliff, be separate rather than public. Despite the efforts of this group, the company established a public school[47] and a separate school was never built in Copper Cliff. Obviously, there was at that time some cohesion among the French Canadians living in the village of Copper Cliff.

Although some French Canadians were employed as carpenters and blacksmiths by the company, most worked in a labouring capacity at the roast yards and in the smelter. Although some cohesion had been apparent during the early days of settlement, the French Canadians did not establish a separate, distinctive settlement. Rather, they lived in small groups in Shantytown, at the East Smelter, in the adjacent Johnson Extension and in Little Italy.[48]

These small settlements survived as extensions of the French-Canadian community in Sudbury, rather than as a cohesive unit within Copper Cliff. With the mass exodus out of town following the post-World War I work reductions, the French-Canadian population moved into Sudbury where there was a French-language separate school system, a French-language Roman Catholic Church, and a vibrant French-Canadian community, much of which was centred in the Flour Mill section of the town. The increased proportion from 1921 to 1931 of Flour Mill residents working for mining companies coincided with the movement of French Canadians out of Copper Cliff.[50]

Table 4.3 Number and Percentage of French Canadians Living in Copper Cliff[49]

Year	Total	French Canadi-ans	Percent of Total
1911	3,082	251	8.14
1921	2, 597	179	6.89
1931	3, 173	284	8.95
1941	3, 732	309	8.27
1951	3, 974	320	8.05
1961	3, 600	374	10.39
1971	4, 100	600	14.63

Despite the stated belief of the French-Canadian community in Sudbury,[51] there is no evidence that company officials in Copper Cliff discriminated against French Canadians any more than they did against any other non-Anglo-Saxon labouring group.[52] The few company-owned houses which were rented to non-Anglo-Saxons were occupied by either Finnish or French-Canadian families. There is no record that other non-Anglo-Saxon elements were ever accorded this privilege.

Poles

Polish immigrants, like those from Finland, were accorded the nationality of the powers which controlled their country in the 1901 Canadian census. Most of the Poles in Copper Cliff had emigrated from the German-controlled section of Poland, and although the term Polish was used in the 1901 census report, the 167 Poles who formed 4.5 percent of the total population of Copper Cliff were counted as Germans.[53] In subsequent census reports, the Polish population of Copper Cliff was correctly designated as Polish.

The Polish community was located in Shantytown, along with the Finnish and French-Canadian communities. Although Polish immigrants had settled near Sudbury in 1884, there is no record of their employment in Copper Cliff until 1889.[54] By the mid-1890s, however, the Polish community was well established in the village and in communication with Polish settlements elsewhere, including those in the Wilno-Barry's Bay area north of Peterborough, Ontario, from whence Polish-speaking priests visited Copper Cliff to provide Polish-language church services.[56] This small community was cohesive enough to build a Roman Catholic church in Shantytown in 1898, and they named it St. Stanislaus Kostka. The church was attended by all Roman Catholics in Copper Cliff and by the growing number of Ukrainian Greek Catholics.[57]

Table 4.4 Number and Percentage of Poles Living in Copper Cliff[55]

Year	Total	Polish	Percent of Total
1911	3,082	126*	4.08
1921	2, 597	48	1.84

Year	Total	Polish	Percent of Total
1931	3, 173	134	4.22
1941	3, 732	58	1.55
1951	3, 974	48	1.21
1961	3, 600	38	1.06
1971	4, 100	15	.37

* There were 72 Germans in Copper Cliff in 1911.

The Polish population, which had never been large, decreased substantially after the turn of the century. A further reduction occurred in response to the decreased industrial activity following World War I. Many of those who left Copper Cliff moved into Sudbury, while others moved to Toronto, Windsor and other centres, to join Polish communities with which Copper Cliff Poles had retained social relations. As economic conditions improved, the Polish population in Copper Cliff began to increase. This was altered, however, by the unemployment of the 1930s and the advent of World War II, both of which resulted in another outward migration, one from which the community never recovered. By 1971 there were only fifteen Poles living in Copper Cliff.

Italians

In 1901 there were approximately 328 Italians living in Copper Cliff, forming almost 9 percent of the population.[58] By the 1911 census, Italians were the second-largest ethnic group in Copper Cliff after the Anglo-Saxons. The Italian community accounted for over 20 percent of the Copper Cliff population, and they remained in that position until 1971 when the proportion decreased.

Table 4.5 Number and Percentage of Italians Living in Copper Cliff[59]

Year	Total	Italian	Percent of Total
1911	3,082	620	20.11
1921	2, 597	541	20.83
1931	3, 173	682	21.49
1941	3, 732	784	21.00
1951	3, 974	825	20.76
1961	3, 600	758	21.06
1971	4, 100	625	15.24

The expansion of company facilities during the late 1890s and early 1900s had attracted an influx of Italian immigrant construction workers, many of whom remained in Copper Cliff to work at the roast yards and smelters. They settled in an area north of the Copper Cliff settlement and as their community expanded it became known as Little Italy. The immigrants, many of whom had emigrated from the same village, leased land from the company, built houses and created a closely knit, cohesive community of people who spoke the same language, had a common religion and shared a culture. In 1903 the CCC president had stated: "An Italian is a bad customer to deal with unless you keep him afraid of you, and that is what we will do."[60] Such discriminatory attitudes, combined with the exclusion of Italians from company-owned recreational facilities, reinforced the alienation imposed by language and culture. The establishment within Little Italy of Italian-language stores and services allowed people in the Italian community to avoid contact with those living in the Anglo-Saxon enclave. In 1914 the community further isolated itself by building an Italian-language Roman Catholic church, St. Elizabeth Church, and abandoning St. Stanislaus, thereby restricting contacts with non-Italian-speaking residents.

In 1914 a Methodist mission was established in the community by Italian-speaking clergy and teachers with the financial backing of the company. Those operating the mission, intent on assimilation, provided kindergarten facilities for children, English-language instruction for adults and a variety of social and educational activities. The Methodist Mission Board leased land within Little Italy from the company and erected a church and rectory.[61] This church, unlike the Finnish Presbyterian mission, was not a divisive influence among the Italian immigrants. Although few members of the Italian community attended the church services, many participated in its educational and social events. The Italian population was either unaware of, or chose to ignore, the assimilative purpose of the mission. When the mission was closed by the United Church of Canada in 1928, its removal caused no dislocation among the residents.[62]

Following the post-1918 industrial reductions, many Italian immigrants left Copper Cliff to move into the Gatchell subdivision, located between Sudbury and Copper Cliff, or into Sudbury itself. The St. Elizabeth Church discontinued operations during the 1920s, and in 1934, with an expanded and still cohesive population, the community remodelled the building as a recreational club. In 1949 members of the community replaced the former church with a new, modern Italian Club which became the focal point for social, cultural and recreational activities in the community.[63]

The aspirations of some residents of Little Italy to join the larger community in Copper Cliff during the 1950s were thwarted by company restrictions and policies. Company houses were not rented to Italians, not even to those born in Canada, and the few houses that the company sold from time to time were not sold to Italians. The purchase of privately owned housing required company consent, and that was not available to Italians who wished to relocate outside Little Italy. As well, the company was not willing to allow the Italian community to expand, and applications from those wishing to build their own homes within Little Italy were denied. This was a prime factor in the stagnation of the community. Young couples, unable either to lease building lots within the Italian community or to settle within Copper Cliff outside Little Italy, moved into Sudbury and became part of the Italian community of that city. As with the Anglo-Saxon community in Copper Cliff, the Italian community aged and was not revitalized.

Ukrainians

The Ukrainians were another of the groups whose national identity was not recognized by the method employed by census enumerators.

Table 4.6 Number and Percentage of Ukrainians Living in Copper Cliff[64]

Year	Total	Ukrainians	Percent of Total
1911	3,082	260*	8.43
1921	2, 597	13**	.50
1931	3, 173	33	1.04
1941	3, 732	51	1.36
1951	3, 974	51	1.28
1961	3, 600	52	1.44
1971	4, 100	50	1.22

Since many of the Ukrainians living in Copper Cliff had emigrated from the Austrian-dominated provinces of the Western Ukraine, they were considered, for census purposes, to be Austrians. This situation existed until the 1921 census. In 1901 there were only twelve Ukrainians (Austrians) living in Copper Cliff, and they formed a mere .29 percent of the total population.[65] Although their numbers and proportions had increased by the time the 1911 census was taken, this was reversed in later censuses.

Ukrainian immigrants had begun arriving in Copper Cliff during the late 1890s, attracted by the possibility of construction jobs, and remained to work as labourers in the smelter.[66] They joined French Canadians already living in the East Smelter settlement and in the Johnson Extension.[67] The Ukrainian community was sufficiently cohesive in 1908 to build a Greek Catholic church in Shantytown on land leased from the company. The company, however, controlled all building on its land and the Greek Catholic congregation was obliged to satisfy company demands concerning the size and style of the church steeple.[68] The Anglo-Saxon population of Copper Cliff was unaware that Poles and Ukrainians were two distinct national groups and referred to both as Poles or "Pollacks." Thus St. Nicholas Church, the Greek Catholic church, became known to Anglo-Saxons in Copper Cliff as the Polish Catholic church.[69]

The Ukrainian population had by 1911 become firmly established in the Johnson Extension where retail outlets, owned by fellow Ukrainians, catered to their needs. One entrepreneur, for example, operated a grocery store, rented houses and sold building lots in the Johnson Extension. He had purchased the unsold lots in the old Johnson subdivision, and had redesigned and re-registered the subdivision under his own name. Ukrainian immigrants moved into the Johnson Extension by choice, both to escape direct company control and to live among people who spoke their language and shared their customs.[70] There is no evidence the company refused to lease land on company-owned property to this group.

The Ukrainian community disappeared from Copper Cliff following World War i. As Austrian nationals, the Ukrainians had been declared enemy aliens by the Canadian government, which registered and interned them under the terms of the 19 October 1914 order-in-council (PC 2721). By the end of the war only twenty Ukrainian families remained in Copper Cliff,[71] and by 1921, with the post-war depression under way, their numbers had decreased to thirteen. Their church was sold and, in 1928, was demolished. A Ukrainian community in Sudbury was begun by those who had left Copper Cliff,[72] and by 1925 the company had purchased all the privately owned property in the Johnson Extension and had added that area to its slag-dump.[73] A few Ukrainian families were still scattered throughout Shantytown, but the Ukrainian community, as an identifiable and cohesive entity, had vanished.

SUMMARY

Obviously neither cohesiveness nor insularity existed among the total population of Copper Cliff. Where it did exist within individual ethnic groups, it was soon eroded, and it is possible that it only existed in the perception of those outside the various ethnic communities. The unwillingness of residents to voice opinions contrary to those of the company gave the myth of cohesion added credence.

CONCLUSION

The images and the realities of life in Copper Cliff are a reflection of life in other company towns, where the images and realities also derived from the control exerted by the owning company over the town, the social landscape and the population. Address indicated status, based on occupation and ethnicity, in most company towns. Companies did make a profit on their housing, although they did not advertise the fact, and companies did export their prospective welfare cases. Few company towns were ever cohesive entities, and the situation in Copper Cliff, where the company imposed spatial segregation on those already isolated by language and culture, was repeated in most Canadian company towns. Despite company control, however, life in a company town was perceived by potential townspeople as being desirable.

Notes

Author's note: The author is indebted to Ron Orasi and the Inco Archives for assistance in the preparation of this article.

[1] For descriptions of the Gorringe and the Copper Cliff Clubs please see Eileen Alice Goltz, "The Exercise of Power in a Company Town: Copper

Cliff, 1886-1980" (Ph.D. thesis, University of Guelph, 1989), 87, 88, 161, 162.

[2] This belief has been enunciated by observers of the dynamics of company towns on both sides of the Atlantic Ocean. Among the Canadian examples are Gilbert A. Stelter, "Community Development in Toronto's Commercial Empire: The Industrial Towns of the Nickel Belt, 1883-1931," *Laurentian University Review* 6 (1974): 13, 14, 41; A.K. Grimmer, "The Development and Operation of a Company-Owned Town," *Engineering Journal* 17 (May 1934): 221-23; Roger A. Roberge, "Resource Towns: The Pulp and Paper Communities," *Canadian Geographical Journal* 94 (1977): 32; Martin Nordegg, *The Possibilities of Canada Are Great: Memoirs, 1906-1924* (Toronto, 1971), 195, 196.

[3] Inco Archives, John D. Evans Diaries, 30 August, 2-9 October, 26-27 October 1886; Copper Cliff rental records; *INCO Triangle,* February 1943, 4.

[4] Inco Archives, Copper Cliff rental records.

[5] Ibid.

[6] For a fuller description of the housing built during this period, please see Eileen Goltz, "A Corporate View of Housing and Community in a Company Town: Copper Cliff, 1886-1920," *Ontario History* 82, no.1 (March 1990): 436.

[7] Inco Archives, Box B0115, Housing Applications, Copper Cliff, op. cit., passim.

[8] Grimmer, "The Development and Operation of a Company-Owned Industrial Town."

[9] Most Canadian writers accept the premise that companies lost money on their towns, as expressed by Harry W. Walker, "Canadian New Towns," *Community Planning Review* 4 (1954): 86. A contrary opinion is expressed in J. Alexander Walker, "Planning of Company Towns in Canada," *Canadian Engineer* 55 (1927): 147.

[10] For British examples, see Sidney Pollard, *The Genesis of Modern Management: A Study of the Industrial Revolution in Great Britain* (London, 1965), 71; William Ashworth, *The Genesis of British Town Planning: A Study in Economics and Social History* (London, 1959), 131. For American examples, see John S. Garner, *The Model Company Town: Urban Design through Private Enterprise in Nineteenth Century New England* (Amherst, Mass., 1984), 218; R. Ely, "Pullman: A Social Study," *Harper's New Monthly Magazine* 70 (February 1885): 99, 101, 105.

[11] Inco Archives, Box 086-44, LB1, letter from John D. Evans, Copper Cliff, to H.P. Mcintosh, Cleveland, 5 September 1890, 650-51; Copper Cliff rental records.

[12] Ibid., LB2, letter from Dr. E.D. Peters, Jr., Copper Cliff, to H.P. Mcintosh, Cleveland, 18 October 1889, 4.

[13] Ibid., Box 090-42, LB3, letter from A.P. Turner, Copper Cliff, to A. Monell, New York, 6 February 1903, 468.

[14] Ibid., LB6, letter from A.D. Miles, Copper Cliff, to John Lawson, Copper Cliff, 30 August 1913, 553.

[15] Ibid., letter from A.D. Miles, Copper Cliff, to W.A. Bostwick, New York, 11 October 1913, 619.

[16] Inco Archives, Copper Cliff rental records; Town of Copper Cliff assessment records.

[17] The subject of profit maximization in Copper Cliff is covered in greater detail in Eileen Goltz, "The Economics of Housing in a Company Town," in *Municipal Development in Northeastern Ontario: Copper Cliff and Sudbury,* edited by Mary Ann Beavis (Winnipeg, 1991), 3-24.

[18] Inco Archives, Copper Cliff rental records; Town of Copper Cliff assessment records.

[19] Ibid.

[20] For a discussion of the image offered of life in company towns see Goltz, "The Exercise of Power in a Company Town," chap.1. American writers have described life in company towns more negatively than have Canadian writers. However, a recent American publication

questions this negative assessment. See Garner, *The Model Company Town,* xi, xii, 3. Few Canadian writers portray company-town life unfavourably. One who does is Rex Lucas, *Minetown, Milltown, Railtown: Life in Canadian Communities of Single Industry* (Toronto: University of Toronto Press, 1971), 14-15, 105, 112, and passim.

[21] Douglas J. Porteous, "The Company Town," in his *Environment and Behaviour* (Reading, Mass., 1977), 331; R. Ely, "Pullman," 465; Richard C. Wade in Stanley Buder, *Pullman: An Experiment in Industrial Order and Community Planning, 1880-1930* (New York, 1965), vii; "Life in a Company Town," *The New Republic* 92 (22 September 1937): 171; Marquis W. Childs, "Company Town," *The Forum and Century* 100 (November 1938): 255.

[22] O.W. Saarinen, "Provincial Land Use Planning Initiatives in the Town of Kapuskasing," *Urban History Review* 10, no.1 (June 1981): 3, 4.

[23] Eileen Goltz, "Espanola: The History of a Pulp and Paper Town," *Laurentian University Review* 6, no.3 (June 1974): 81.

[24] Inco Archives, Copper Cliff rental records; Town of Copper Cliff assessment records.

[25] Inco Archives, Box BOI-45, Housing Applications, Copper Cliff, 1926—42, 1943-48, 1949-51, 1952-56, 1957-59, 1960-66, 1967-69, 1970-73, other (the dates refer to the year

the applicant commenced employment with Inco), passim.

[26] J.M.S. Careless, "Frontierism, Metropolitanism and Canadian History," *Canadian Historical Review* 34, no.1 (March 1954): 1-21.

[27] The earliest ethnic data available is from the 1901 Canadian census, taken just prior to the incorporation of Copper Cliff as a town. This census combined the populations of McKim and Snider townships (within which the village of Copper Cliff was located) with that of the sparsely populated Waters Township. Most of the population of the combined townships resided in Copper Cliff, and although complete statistical accuracy is impossible, the data does provide an indication of the town's ethnic mix at that time.

[28] *Census of Canada,* 1911, 1921, 1931, 1941, 1951, 1961, 1971.

[29] Ibid., 1901.

[30] Inco Archives, Box 090-42, LB6, letter from A.P. Turner, Copper Cliff, to A. Monell, New York, 25 November 1912, 381; letter from A.P. Turner, Copper Cliff, to James Ashley, New York, 11 December 1912, 425; Copper Cliff rental records; Town of Copper Cliff assessment records; Debenture Register, Bylaw 18, 1900 (Township of McKim Bylaws), 2; Copper Cliff School Board Minutes, 30 June 1900; *Vernon's Sudbury and Copper Cliff Directory;*

Sudbury Journal, 25 April, 16 May, 3 October 1901.

[31] Inco Archives, Copper Cliff rental records; Town of Copper Cliff assessment records; *The Church of England in the City and District of Sudbury, Ontario* (Sudbury, 1936), 16; *Sudbury Journal,* 23 June 1892, 17 June, 29 September, 27 October, 1898, 4 May, 25 May, 31 August, 19 October, 23 November, 21 December 1899, 30 December 1900, 14 July 1910, 22 June, 20 July, 30 November 1911.

[32] Inco Archives, Box 090-42, LB2, letter from A.P. Turner, Copper Cliff, to A. Monell, New York, 19 January 1903, 387; Box 090-42, LB3, letter from A.P. Turner, Copper Cliff, to A. Monell, New York, 19 January 1903, 387-88; Copper Cliff Inventory, circa 1913-15, 406; letter from L. Lamb, Bank of Toronto, Toronto, to Frederick P. Bernhard, St. Petersburg, Florida, 24 February 1953; Box 090-42, LB6, letter from A.D. Miles, Copper Cliff, to John Lawson, Copper Cliff, 30 August 1913, 553; Copper Cliff rental records; Town of Copper Cliff assessment records; *Sudbury Journal,* 11 April, 16 May, 31 October, 5 December 1901; 27 May 1902; 11 February, 28 April 1904. *Copper Cliff Courier,* Christmas Number (1903): 10; *Sudbury Star,* 2 October 1915, 4; 10 May 1916, 4. The information concerning the exclusion of non-Anglo-Saxons from the Copper Cliff

Club was obtained from members of both the Anglo-Saxon and the Italian communities.

[33] Inco Archives, Copper Cliff rental records; Town of Copper Cliff assessment records; F.B. Howard-White, *Nickel: An Historical Review* (New York, 1963), 214; Alice O. Riley, *Chronology of Nickel,* prepared for Inco internal use [1984], 158, 159; *Sudbury Star,* 1 December 1920, 4; 24 September 1921, 2, 4; 15 October 1921, 4; 12 November 1921, 2; 16 November 1921, 2; 19 November 1921, 2; 23 November 1921, 2; 30 November 1921, 1; 7 December 1921, regular edition, 5; 14 December 1921, 3; 2 January 1922, 3; 11 January 1922, 2, 3; 18 January 1922, 5; 28 January 1922, 5; 1 February 1922, 2, 3; 31 May 1922, 2, 3; 17 June 1922, 5; 12 August 1922, 3; 16 August 1922, 5; 19 August 1922, 3, 7; 23 August 1922, 3; 26 August 1922, 4, 8; 30 August 1922, 8.

[34] Inco Archives, Box 092-40, LB2, letter from John Lawson, Copper Cliff, to W.A. Bostwick, New York, 2 April 1913, 324; Box 093-42, LB2, letter from A.D. Miles, Copper Cliff, to A. Monell, New York, 25 April 1914, 27.

[35] Inco Archives, Copper Cliff rental records; Town of Copper Cliff assessment records; *Sudbury Star,* 6 May 1925, 6; 3 June 1925, 11; 1 August 1925, 1; 16 September 1925, 3; 26 September 1925, 10; 3 October 1925, 8; 23

December 1925, 3; 17 March 1926, 10; 10 April 1926, 11; 8 May 1926, 11; 15 May 1926, 12; 12 July 1927, 6.

[36] Inco Archives, Nickel Basin Properties Limited, files: a series of memos from H.M. Stephenson, Copper Cliff, to R.L. Beattie, Copper Cliff, written in September 1933 (re: sale of Nickel Street houses in Copper Cliff); May, June, July, August 1934 (re: sale of Copper Cliff houses); a series of memos from H.M. Stephenson, Copper Cliff, to Charles Dorion, Copper Cliff, written in October 1935, May, August, September 1936 (re: sale of Copper Cliff houses); memo from H.M. Stephenson, Copper Cliff, to J.R. O'Donnell, Copper Cliff, 10 September 1937 (re: sale of building at 21 Rink Street [Smelter Club in Copper Cliff]); Copper Cliff rental records; Town of Copper Cliff assessment records; *Sudbury Star,* 19 July 1933, 3; 26 May 1934, 3.

[37] *Census of Canada,* 1901.

[38] Ibid., 1911, 1921, 1931, 1941.

[39] Inco Archives, "Recollections of Mrs. A.H. Hibbard, whose father was Mr. Louis [*sic*] Ashmun, first superintendent of Sudbury Mines," 2, 3; Martha Isobel Gerrard Allen, "A Survey of Finnish Cultural, Economic and Political Development in the Sudbury District of Ontario" (M.A. thesis, University of Western Ontario, 1954), 47, 81; Mauri A. Jalava, "The Finnish

Community in Copper Cliff, 1885-1915" (Paper presented to the Sudbury and District Historical Society, 1980), 6, 9, 10; *Sudbury Journal,* 4, 11 July 1895; *Sudbury Star,* 2 July 1932, 5.

[40] Inco Archives, Box 093-42, LB1, letter from A.D. Miles, Copper Cliff, to W.A. Bostwick, New York, 12 February 1915, 113; Town of Copper Cliff Council Minutes, 2 June 1903; Allen, "A Survey of Finnish Cultural, Economic and Political Development in the Sudbury District of Ontario," 2324; Jalava, "The Finnish Community in Copper Cliff, 1885-1915," 9-16; *Copper Cliff Courier,* Christmas Number (1903): 16; *Sudbury Journal,* 15 January 1915; *Sudbury Star,* 8 March 1913, 8; 24 May 1913, 8; 28 June 1913, 1; 9 August 1913, 2; 1 April 1914, 1; 17 June 1914, 6; 5 December 1914, 6; 13 February 1915, 4; 23 February 1915, 4; 24 March 1915, 4; 1 April 1916, 4; 20 January 1917, 4; 13 February 1918, 4; 22 May 1918, 8; 8 February 1919, 4; 26 April 1921, 4; 5 October 1921, 2; 2 November 1921, 10; 29 April 1922, 3; 16 September 1922, 2; *Vernon's Copper Cliff and Sudbury Directory,* 1916.

[41] Allen, "A Survey of Finnish Cultural, Economic and Political Development in the Sudbury District of Ontario," 33, 34, 47, 81; Jalava, "The Finnish Community in Copper Cliff, 1885-1915," 9-16; *Sudbury Journal,* 4, 11 July

1895; *Vernon's Sudbury and Copper Cliff Directory,* 1911, 1914, 1916, 1918-19, 1921-22.

[42] Inco Archives, Copper Cliff rental records; *Sudbury Star,* 28 October 1914, 4; 5 December 1914, 6; 9 January 1915, 4; 17 April 1918, 4; *Sudbury Journal, 9* May 1918.

[43] Inco Archives, Nickel Basin Properties Limited files, letters from Rev. J.W.E. Newbery, All Peoples' Mission, Sudbury, to E.A. Collins, Copper Cliff, 24 June 1938 (enclosed: "Names of the Finnish United Church People in Copper Cliff"), 30 June 1938; letter from E.A. Collins, Copper Cliff, to Rev. J.W.E. Newbery, All Peoples' Mission, Sudbury, 5 July 1938; application for a lot for a Church from All Peoples' Mission, 30 June 1938; Copper Cliff rental records; Town of Copper Cliff assessment records; Charles Dorion, *The First 75 Years: A Headline History of Sudbury, Canada* (Ilfracombe, Devon, 1958), 36; *Sudbury Star,* 20 July 1938, 3.

[44] *Census of Canada,* 1901.

[45] Both the Town of Copper Cliff assessment records and the Inco rental records have been studied to reach this conclusion.

[46] Town of Copper Cliff assessment records.

[47] Inco Archives, Box 086-44, LB1, letter from John D. Evans, Copper Cliff, to H.P. Mcintosh, Cleveland, 6 November 1889, 92; 9 January 1890, 258-60; letter from Dr. E.D. Peters, Jr.,

Copper Cliff, to H.P. Mcintosh, Cleveland, 22 October 1889, 20; 1 December 1889, 165-76; *Canada Census,* 1891, manuscript.

[48] Gail Cuthbert Brandt, "The Development of French-Canadian Cultural Institutions in Sudbury, Ontario, 1883-1920," *Laurentian University Review* 11, no.2 (February 1979): 7-8; D.H. Browne, quoted in D.M. LeBourdais, *Sudbury Basin: The Story of Nickel* (Toronto, 1953), 53; *Sudbury Star,* 15 April 1914, 15; 6 December 1916, 6.

[49] *Census of Canada,* 1911, 1921, 1931, 1941, 1951, 1961, 1971.

[50] Stelter, "Community Development in Toronto's Commercial Empire," 7.

[51] This belief has been expressed to me by several members of the French-Canadian community in Sudbury. One of these is Dr. Gaetan Gervais, Department of History, Laurentian University, Sudbury.

[52] Inco Archives, Copper Cliff rental records; Town of Copper Cliff, assessment records; *Town of Copper Cliff Voters Lists,* 1903, 1904, 1907, 1908, 1914, 1917, 1918, 1924.

[53] *Census of Canada,* 1901.

[54] Inco Archives, Z07-789, Cash Books and Ledgers, Ledger B, 1889-1990, 094-14-3, 621; Cash Book B, 1887-89, 094-26-1, 14 October 1889, 164; Henry Radecki "Polish Immigrants in Sudbury, Ontario" (Paper presented at the

conference "Poles in North America," Toronto, 1980), 7; Florence Howey, *Pioneering on the CPR* (privately published, 1938), 110-11 (the book was reprinted in Sudbury in 1968 under the title *Sudbury Minus One*).

[55] *Census of Canada,* 1901, 1911, 1921, 1931, 1941, 1951, 1961, 1971.

[56] Radecki, "Polish Immigrants in Sudbury, Ontario," 2, 3, 10; Jozef Samulski, *Pamientnik Emigranta Polskiego w Kanaadzie* (Wroctow, Poland, 1978), 45 (translated by Dr. Henry Radecki); *Sudbury Journal,* 29 October 1896.

[57] Inco Archives, Copper Cliff rental records; Town of Copper Cliff assessment records; *Sudbury Journal, 29* September, 27 October 1898.

[58] *Census of Canada,* 1901.

[59] Ibid., 1911, 1921, 1931, 1941, 1951, 1961, 1971.

[60] Inco Archives, Box 090-42, LB3, letter from A.P. Turner, Copper Cliff, to E.F. Wood, New York, 9 May 1903, 843.

[61] Inco Archives, Copper Cliff rental records.

[62] *Sudbury Star,* 9 December 1914, 5; 23 May 1916, 4; 7 July 1917, 4.

[63] Inco Archives, Copper Cliff rental records; Copper Cliff Museum, *A Bit of the Cliff,* edited by Rae Swan and Margaret Bertulli (1982), 60, 61; *Sudbury Star,* 3 October 1928, 1; Conversations with several residents from Little Italy in Copper Cliff.

[64] *Census of Canada,* 1911, 1921, 1931, 1941, 1951, 1961, 1971.

[65] Ibid., 1901.

[66] Ibid., 1901.

[67] Samulski, *Pamientnik Emigranta Polskiego w Kanaadzie, 47.*

[68] Inco Archives, Box 090-42, LB2, letters from A.P. Turner, Copper Cliff, to Andrew Ostroski, Copper Cliff, 27 June 1908, 685; 3 July 1908, 712.

[69] *Vernon's Sudbury and Copper Cliff Directory,* 1911, 1914, 1916, 1918-19, 1921-22; A.P. Coleman, "Ontario Mines and Miners," in *Empire Club of Canada Addresses, 91-1912* (Toronto, 1913), 164; M.H. Marunchak, *The Ukrainian Canadians: A History* (Winnipeg, 1970), 215; *Sudbury Journal,* 13 February, 18 June 1908, 28 April 1910; *Sudbury Star,* 8 November 1913, 1.

[70] Land Titles office, Sudbury, M Plans 54 and 55 (these are available on microfilm at the Laurentian University Library, Sudbury); Inco Archives, Copper Cliff rental records; Town of Copper Cliff assessment records; *Town of Copper Cliff Voters' List,* 1914, 1917, 1918.

[71] Mary Stefura, "Sudbury Ukrainian Time Line," in L.Y. Luciuk, and I.L. Wynnyckyj, eds., *Ukrainians in Ontario, Polyphony* 10, double issue (1988): 67.

[72] Town of Copper Cliff assessment records; City of Sudbury assessment records; Marunchak, *The Ukrainian Canadians,* 215, 216; *Census of Canada,* 1911, 1921.

[73] Inco Archives, Copper Cliff rental records; Town of Copper Cliff assessment records; Town of Copper Cliff correspondence, 1923-24, "Treasurer's Sale of Land for Taxes," 16 June 1932; "Notice" of Sale, 30 June 1924; Town of Copper Cliff, Bylaw No.122, 26 November 1923; *Sudbury Star,* 21 April 1923, 4.

Part Three

The Era of Government Intervention

This picture was taken in Copper Cliff in 1893, when concerns about mining safety were minimal.

Introduction

Even in the early years, government policies significantly affected the establishment of mining communities. As Wallace has reminded us, were it not for government-subsidized and government-built railroads that criss-crossed vast northern distances, most mining communities would never have come into being. And, as Bray has pointed out, once established, mining companies such as the Canadian Copper Company (CCC) and mining communities such as Sudbury often lobbied provincial and federal politicians, with mixed success, for beneficial legislation.

In the post-World War II era, however, government increasingly took the initiative on issues related to mining and mining communities in Northern Ontario. In part, this may simply have been a by-product of the growth of government activities generally during these years. Or, as Robert Robson suggests below, it could have been the result of a recognition by governments of the revenue-generating potential of the area. Whatever the reasons, in terms of government intervention, the 1950s represented a watershed in the relationship between mining companies and their communities.

The three papers presented here take very different positions on the central question of whether this heightened governmental intervention was either desirable or efficacious. In "Building Resource Towns: A

Decade of Government Intervention," Robert Robson, from the University of Manitoba, acknowledges that during the 1950s the government of Ontario was at times as non-interventionist as it had been in an earlier era, but he also points out some of the direct steps taken by the Frost administration to shape the development of mining communities. Manitouwadge, for example, stands out as a planned community destined to serve as a model for many others. Robson, however, concludes that the failure of the Ontario government to develop a "comprehensive policy statement" defining its interventionist objectives "exacerbated an already difficult situation and furthered the confused state of affairs in the resource sector."

Sometimes governments were spurred to action because of immediate crises in the mining industry. In his paper, "Breakthroughs in Health and Safety in Northern Ontario Mines," Laurentian's Donald Trotter reviews four major inquiries set up in response to urgent community pressure to examine the health and safety of miners. Trotter optimistically concludes that "the actions taken ... are giant steps in the right direction. Ontario's mining industry presently has an enviable safety record." On the other hand, in "Government Regulation of Health Hazards in the Ontario Uranium Mining Industry, 1955-1976," University of Toronto sociologist Brian Walker draws a contradictory conclusion. For Walker, government regulation of the uranium industry in Elliot Lake did not reflect the concerns of the community of miners but rather those of the uranium industry itself. According to his interpretation, health and safety were given less priority than national security and economic considerations.

Chapter 5

Building Resource Towns: Government Intervention in Ontario in the 1950s

Robert Robson

From its formation in 1954 until approximately I960, the Cabinet Committee on Townsites coordinated resource-town development in the province of Ontario. Mandated to "formulate" and to "carry out policy" with respect to "new towns," the committee was charged with the responsibility of overseeing the new town phenomenon as it evolved in Ontario.[1] This included not only the supervision of community development, but also, and perhaps more importantly, policy formulation. While town plans were plotted, townsites cleared and services installed, it was particularly at this more theoretical level that the cabinet committee toiled with the nuances of the resource town. For the most part, it did so with a great deal of difficulty. Although the cabinet committee initially responded to the resource-town challenge with unbridled enthusiasm, by 1960 it had become more and more reluctant to become involved in the resource-town process. Nevertheless, during this six-year period, the committee's activity followed four rather vague policy objectives. Ranging from the non-interventionist stance of a previous era through to the complete orchestration of community development, government policy as defined by the cabinet committee reflected the uncertainty of the resource production process.

The willingness of the provincial government to assume a more active leadership role in the resource-town phenomenon was characteristic of the post-war era of northern expansion.[2] Recognizing the value of the untapped resources of the North, the provincial government became an active participant in a variety of northern development schemes. These initiatives, which ranged from road-building to the provision of community infrastructure, manifested a philosophy of development at all costs. At the same time, the various programs introduced by the provincial government also clearly reflected the sovereignty issues of the post-war period, a concern for northern stability, the general acceptance of the concept of regional planning, the increasingly competitive labour market and the prevailing negative attitudes towards the "company town" of old. In the end, then, the provincial government, through its efforts to rationalize northern growth and expansion, committed itself to the notion of comprehensive planning.

At the community level, comprehensive planning meant that the provincial government attempted to encourage the incorporation of comprehensive social and economic principles into the fabric of physical planning. When applied, the comprehensive initiative gave rise to land-use segregation, greenbelt areas, the separation of vehicular and pedestrian traffic, environmental management and in some cases, the incorporation of the neighbourhood unit concept. Manitouwadge, Elliot Lake and Cardiff were all comprehensively planned resource towns. The wide winding streets, the intertwining neighbourhood units, the segregation of the community by function, and the separation of pedestrian from vehicular traffic were, to one degree or another, incorporated into all three communities. At the same time, because of an ongoing policy conundrum, the provincial government also on occasion discouraged community growth, concentrated community activity in selected growth centres, and/or simply allowed the resource company to devise its own community development program. While so-called model towns were indeed being created as a result of the comprehensive planning initiative, government policy concerning new resource towns was still very much in a state of confusion. For the most part it constituted little more than pragmatic responses to particular situations. Although government intervention was certainly in vogue during the tenure of the Cabinet Committee on Townsites, it was exercised selectively, with little consideration given to the standardization of policy and procedures.

THE ERA OF GOVERNMENT INTERVENTION

In the fall of 1954, the provincial government established the Cabinet Committee on Townsites. Composed of the minister of municipal affairs, G.H. Dunbar; the minister of mines, P.T. Kelly; the minister of lands and forests, C.E. Mapledoram; and chaired by W.T. Warrender, minister of planning and development, the committee was intended to meet the municipal needs of communities located beyond the settled districts of the province.[3] In effect, it was the first provincial government agency designed singularly to accommodate the unique settlement needs of the resource community. While the cabinet committee concentrated its efforts on policy formulation, an administrative subcommittee was established to coordinate the committee's field work. Composed of F.W. Beatty, surveyor-general in the Department of Lands and Forests; J.W.P. Carter, director of the General Administration Branch in the Department of Municipal Affairs; H.C. Rickaby, deputy minister of mines; D.F. Taylor, chief planner in the Department of Planning and Development; and chaired by A.E.K. Bunnell of the Department of Planning and Development, the subcommittee was a working committee charged with responsibility for establishing "townsite need, location and population potential, both present and future."[4] Although the subcommittee was created as the investigating appendage of the cabinet committee, much of its responsibility devolved to the Department of Planning and Development. Indeed, one of the subcommittee's earliest recommendations indicated that "the Department of Planning and Development, which had a permanent staff of competent planning personnel, would be responsible for townsite location, basic surveys and design of all townsites in which the government participated."[5] Also involved in the investigative process were the Department of Health on issues related to water and sewer facilities, the Department of Education on matters of school accommodation, and the Department of Highways on questions of transportation.

While the townsite committee attempted to coordinate resource-town activity, the provincial government's position on the creation of resource towns was seemingly fourfold: it preferred to direct community development towards existing urban centres; it pursued a policy of "non-intervention" in the case of a single-resource-company development scheme; it was willing to participate in community development in the event of land-use conflict; and it was prepared to coordinate community development when two or more resource companies were involved in a resource-development scheme. Although the province had created the bureaucratic machinery to contend with resource-community issues, its policy regarding the development of resource towns was confusing and at times even contradictory.

DEVELOPMENT OF EXISTING URBAN CENTRES

The preference of the provincial government for the expansion of existing urban centres was in part a reaction to previous patterns of community growth in the resource sector. The small scattered settlements which had traditionally been associated with the resource industries were described in 1954 by Community Planning officials as "inefficient and undesirable." D.F. Taylor further maintained that this form of community development led to the "creation of communities poorly placed and too small to economically provide the recreational, educational and cultural facilities necessary for the full life."[6]

In an attempt to alleviate these problems, the cabinet committee made a concerted effort to discourage this form of expansion. Primarily because existing communities offered established retail services, schools, sewer systems and water facilities, the provincial government directed new population growth into established centres.

In 1955-56, the provincial government preference for developing existing centres was clearly demonstrated at Sioux Lookout, Cutler and Moose Mountain. In August of 1955 the cabinet committee had received correspondence from the Anglo-Newfoundland Development Company pertaining to the establishment of a pulp and newsprint mill at Pelican Siding – approximately nine kilometres west of Sioux Lookout.[7] The company president estimated that the $50-million facility would employ approximately 500, and would increase the total local population by anywhere from 2, 500 to 4, 500 people. While all housing possibilities had apparently been considered by Anglo-Newfoundland, it initially inquired into the feasibility of erecting a townsite at Pelican Siding. Government policy, as enunciated by the cabinet committee, was completely at odds with the creation of a separate townsite to serve the pulp and paper company. At a meeting on 23 August, the government persuaded Anglo-Newfoundland to focus its community endeavours in the small village of Sioux Lookout.

A similar arrangement was made for Noranda's sulphuric acid operation at Cutler in 1956.[8] By agreement with the administrative subcommittee, Noranda made the village of Spanish, some ten kilometres east of its processing facility, the home community for its aniticipated workforce of from 100 to 150 men.

A final example, and one which is perhaps a better illustration of the central district theme, concerned the Moose Mountain operations of the Hanna Coal and Ore Corporation.[9] With a view to developing the iron deposits in the Moose Mountain area, approximately sixty kilometres north of Sudbury, Hanna personnel approached the administrative subcommittee on townsites to discuss the feasibility of locating community facilities at the mine site. In discussions with the deputy minister of mines, H.D. Rickaby, and E.A. Gomme of the Department of Planning and Development, it was estimated that the ironworks would employ 230 men. Taking into consideration the regional complexion of the Sudbury district, a joint agreement was reached whereby Hanna employees would reside in existing communities in the Sudbury area. Capreol, located forty kilometres south of Moose Mountain, became the main home for the iron workers, although others would take up residence in various towns throughout the region.

In their efforts to encourage the expansion of existing communities, government officials clearly attempted to deal with what they perceived to be the "roots" of the northern settlement problem. The difficult terrain, the absence of local government, the widespread physical distribution of workplaces, the unstable nature of northern industry, and what was cited by the Bureau of Municipal Research as the "almost complete lack of common sense" in the community development process, all contributed to an increasing acceptance of the government's growth-centre approach.[10] Although it was only occasionally implemented during the 1954-60 era, the growth-centre approach eventually came to form the basis of the provincial government's regional development strategy of the late 1960s.

NON-INTERVENTION

In what appeared to be a further effort towards the rationalization of urban development in the North, the townsite committee applied a philosophy of "non-intervention" in areas where it saw no immediate need for government involvement. In situations where a community was already developing on the basis of the activity of a single-resource company, the provincial government adopted a laissez-faire attitude. Offering advice "in a planning consultant capacity," the provincial government extended "assistance as required" but generally allowed the resource company the privilege of developing townsite facilities. By retaining a certain measure of control through the inspection of townsite plans, the province attempted to implement what it described as a "sound development policy."[11]

This non-interventionist policy was applied at least twice during the period: at the Copper Corporation's new mine north of Sault Ste. Marie and at that of Coldstream Copper Mine Limited in the Lakehead region. The provincial government's involvement in the Copper Corporation situation, apparently initiated in 1956 when the company inquired into the possibility of government assistance in the establishment of a townsite, was strictly supervisory. In maintaining that the mineral field was to be developed "on a one mine basis," the cabinet committee allocated townsite responsibility to the Copper Corporation with the provision that the province would intervene only when necessary "to assure high standards of physical development."[12]

The Coldstream development, like that of the Copper Corporation, was based on the activity of a single mining company. In this case, Coldstream personnel approached the townsite committee in March of 1956 with a proposal outlining their mining project.[13] Anticipating a workforce of approximately 275 men, the company stressed the need for a "planned townsite" and further indicated that it intended to erect a number of company-owned houses at the mine site. The proposal also alluded to the participation of several other mining companies in the townsite scheme. It was seemingly on the basis of multi-company involvement that the cabinet committee authorized the allocation of $50,000 for the preliminary townsite investigations.

The townsite studies, conducted by F.H. Deeks of the Community Planning Branch, indicated that there were three possible sites suitable to accommodate the anticipated population of 1, 400.[14] In the interim, however, the several mining schemes referred to in the initial Coldstream proposal had been reduced to two: Coldstream and the Jellicoe Mining Company. By the time the initial townsite studies were completed in October 1957, Coldstream remained the only interested party. As a result of this "disappointing" turn of events, the province decided to pursue a policy which was described by the subcommittee as one of "non-intervention." On this basis, Coldstream was "permitted to establish a small community commensurate with its own needs and wholly within its own property ... subject to the approval of the Department of Planning and Development."[15] In consultation with the Community Planning Branch, the company eventually erected a small community near Burchell Lake, adjacent to the mine site.

Other examples of the non-interventionist theme, such as the Faraday Mine townsite north of Bancroft or the Pronto Mine townsite south of Elliot Lake, further demonstrated the provincial government's attempt to maintain control of townsite affairs through an advisory role.[16] This approach was only moderately successful, as often company objectives did not correspond to government prerogatives. The Faraday townsite, for example, which provided for ribbon development along the Bancroft highway, was severely criticized by the cabinet committee for its haphazard form of growth.[17] In this and other areas, the provincial government's supervisory role was an ineffective method of ensuring responsible townsite development.

CASES OF LAND-USE CONFLICT

Different circumstances often prompted the townsite committee to become a willing participant in resource-town development. In situations where competing land-development schemes clashed, the cabinet committee was quick to intervene. This was particularly true in confrontations over resource land use and recreational land use in Northern Ontario. When such disputes occurred, the cabinet committee and its administrative subcommittee were quite prepared to exercise their authority in townsite affairs in order to facilitate responsible land use.

The best example of government involvement in the issue of northern land-use conflict occurred at Temagami. The village of Temagami was established in 1924 by the Temagami Timber Company. Located adjacent to the Temagami Forest Reserve, the community also catered to the district's tourist trade. The two industries coexisted as the central focus of the area's economy until the mid-1950s. By 1956, however, the development of several nearby mine sites created a potential land-use conflict. Quick to recognize the danger, P.T. Kelly, minister of mines, advised a joint meeting of the cabinet committee and its administrative subcommittee in March of 1956, stating

> that because the area in question was located within one of the most popular and attractive recreational parts of the province, the government townsite committee should work with the mines to assure that the townsite to be created was appropriately located and that conflict between the mining industry and the recreational potential of the area be minimized.[18]

In an effort to minimize the potential conflict between the two industries, the townsite committee authorized the investigation of the district's community needs. Its findings determined that while the Temagami Mining Company sought to establish a separate townsite adjacent to its mine, most local groups favoured the expansion of existing facilities within the community of Temagami. One local group in particular, the Temagami Lakes Association, was vehemently against the establishment of any form of processing facility related to the extractive industry. Indeed, in speaking on behalf of the tourist industry, the Temagami Lakes Association maintained that the mining enterprise would "highly disfigure the scenic beauty of the Lake."[19]

With an anticipated population increase of 2, 500, the company had initiated preliminary townsite surveys in 1955. The final report prepared by an independent consulting firm completely rejected the proposal for expanding the existing community. Entitled "Appreciation of Temagami Village," the study cited the town's poor water supply, inadequate sewage system, narrow confines and "barren rock" surroundings as factors which made expansion impractical. It also mentioned the "great inconvenience to the present cottage owners and tourists" that expansion would entail.[20]

On 8 May 1956, after further investigation, the mining company selected a location for its proposed townsite approximately seven kilometres from the established town. Under the guidance of the cabinet committee, the mining company proceeded to survey the townsite property, making allowances for roadways, parking areas, shopping districts and recreational areas. The survey, however, was as far as townsite development went. The mining company, apparently in an eleventh-hour decision, decided not to establish a "company town," and indeed opted for the expansion of existing facilities. Whatever the reasoning, the company town proved largely unnecessary as "most of the employees engaged by the mine were drawn from existing settlements of the region."[21] In recognizing the difficulties involved in balancing mining activity with the "tremendously important tourist and summer resort values of the area," the Temagami Mining Company combined with the provincial government to facilitate the "moulding of the new and older economic basis" in the one community.[22]

TOWNSITES WITH MORE THAN ONE RESOURCE COMPANY

While the provincial government's role in the Temagami, Burchell Lake or Sioux Lookout townsites was largely supervisory, in other areas its involvement was more direct. Where townsite facilities were deemed necessary to accommodate the needs of two or more resource companies, the province readily accepted responsibility for townsite affairs. Whether the townsite committee was actually approached by interested resource companies or whether the province initiated townsite discussions, the willingness of the provincial government to become involved in such schemes was directly related to the viability of the community. In the opinion of the cabinet committee, the potential for permanent settlement was dramatically enhanced when industrial activity was sponsored by more than one resource company. In these cases, the government supported permanent townsites. When combined with the growing reluctance on the part of resource companies to invest capital in community projects, this led to a much greater participation of the provincial government in townsite development. Manitouwadge, Elliot Lake and Cardiff are vivid examples of this phenomenon.

Manitouwadge

The community of Manitouwadge is situated on the south side of Manitouwadge Lake, approximately 400 kilometres northeast of Thunder Bay and 80 kilometres inland from Lake Superior. While the initial discovery of copper-ore at the site was apparently made in 1931 by a Department of Mines field survey, it was not until 1953 that development work actually began. The original claim, sold by local prospectors to the General Engineering Company, eventually gave rise to the region's major ore producer, the Geco Mine. Several lots adjacent to the Geco property were also consolidated to form the Willroy Mine. At the request of these resource industries, the Ontario government agreed in the spring of 1954 to develop the townsite of Manitouwadge. Cited as the "pilot town of this new field of governmental planning responsibility," the community was the first example of total new town planning and development as undertaken by the cabinet committee and its administrative subcommittee.[23] Manitouwadge is generally recognized as the prototype of Ontario's government-designed resource communities. Indeed, in 1957 a *Financial Post* article provocatively subtitled "Manitouwadge May Set Pattern for Future Towns in the Wilds" suggested: "On this town rests the hope that Canadians with the pioneer spirit will in the future be able to work and play in the wilderness under social conditions as good as their city cousins enjoy, and perhaps better."[24]

In June of 1954 the provincial government initiated preliminary investigations of the Geco townsite proposal. An exploratory excursion was undertaken to determine the feasibility of townsite development. The government "team" that ventured into the mineral field included representatives from the departments of Mines, Lands and Forests, Planning and Development, Highways and Health. Besides the mining company personnel, other interested parties included pulp and paper company officials and Canadian National and Canadian Pacific Railway personnel. Both the Ontario Pulp and Paper Company and Marathon Paper Mills of Canada advised the provincial government that they too would be interested in establishing employee housing within the community.[25]

As the townsite committee struggled with the novelty of Manitouwadge, they also sought to develop a well-founded new town policy. The cabinet committee attempted to capitalize on the experience offered by firms such as Marathon Paper in the development of small communities, and also solicited outside advice. On at least two occasions, for example, D.F. Taylor contacted American firms experienced in resource-town development; on at least one other, he conferred with a Quebec deputy minister whose department had "considerable experience" in townsite development.[26] On a wide variety of questions ranging from lot-disposal methods to the relationship between employee and community population figures, the townsite committee showed an eagerness to formulate policies to develop the townsite according to rational planning procedures.

The formative "observation and recommendation" on the Manitouwadge townsite issue are found in a cabinet committee memorandum dated 22 June 1954. Based on the June investigation as well as on a series of reports filed by Geco personnel, it was determined that

1) financing to the extent of $15,000,000 was now being arranged for the development of the mine, and from examination by the party headed by Mr. Rickaby it would appear that there is a very sizeable body of copper, pyrite and zinc ... It is contemplated that there will be a working force of approximately 1, 200 engaged in the construc-

tion, which in 1956 will reduce to an operating force of approximately 700.

2) A permanent working force of 700 people will demand a townsite to accommodate at least 3, 500 persons. The property being 43 miles north of the C.P.R. and 24 miles south of the C.N.R., it will have to be a self-contained townsite.

3) From the plans and maps which were made available to us, and from our examination on the ground, we are satisfied that a suitable location for a townsite exists on the south shore of Lake Manitouwadge with its easterly end approximately one mile from the Geco ore-body.[27]

As the findings indicate, a favourable appraisal of the townsite was submitted by government officials. This perspective was further emphasized in the administrative subcommittee's recommendations, which in part read:

1) As the Company intends to erect approximately 30 homes this fall, and as no townsite for 3, 500 people should be established in unorganized territory, we reasoned that an area of approximately 12 miles from north to south and 12 miles from east to west and centered in the proposed Manitouwadge townsite be established on the initiative of the Minister of Municipal Affairs, as an Improvement District.

2) That three trustees be appointed on a temporary or interim basis in the persons of Messrs. J.A. Graham, Geco Mine Manager, D.F. Taylor, Depart-

ment of Planning and Development, and W.F. Weaver, Department of Lands and Forests.

3) That the trustees be authorized to engage a consulting engineer to examine the matter of water supply and subsurface conditions.
4) That the land required for townsite purposes which is now vested in the crown be transferred to the Improvement District.
5) That the by-laws in respect of subdivision control and uses of land be passed by the trustees, and regulations be instituted by the appropriate department of Government which will effectively prevent the erection of shacks along the highways and railways leading to the Improvement District.
6) That survey and design of the townsite be placed in the hands of the Departments of Lands and Forests and Planning and Development respectively. That for this purpose a sum of $50,000 be placed at their disposal by the Government. This is to be a charge against the Improvement District recoverable as townsite lots are sold.[28]

By the end of June 1954, the cabinet committee was committed to the development of the townsite. While specific planning, building and administrative problems were still to be overcome, by the end of September the general pattern of the community had been ascertained.[29] The total area of the townsite, which amounted to 648 hectares, was to be developed in three units or neighbourhood sections. The first unit, Neighbourhood Area No.1, and the third unit, Neighbourhood Area No.3, were each intended to accommodate 2, 540 persons on 57 hectares of land. Neighbourhood Area No.2, or the central unit, consisted of 35 hectares and was scheduled to house 1, 740 persons. The total townsite area was rather vividly described as consisting "of three giant blobs of housing development with shopping facilities in the centre. Scattered between and within these areas are park facilities which include rock outcrops and muskeg unsuitable as building land."[30]

The central thrust of the planning initiative appeared to be directed towards the core area. While neighbourhood units contained a few shops, churches and perhaps a community hall, each was planned to allow quick and easy access to the town centre. The core area itself was developed on a ridge of high land immediately adjacent to the southwestern corner of Lake Manitouwadge. In discussing the benefits of such a location, Community Planning Branch's R.N. Percival detailed four major factors:

1) It commands excellent views over the lake and partly over the surrounding development.
2) It has fairly good sub-soil conditions for development.
3) It can be easily integrated with the existing mine road and first stage of development.
4) Traditionally the church, market place and public buildings occupy the highest point in the town.[31]

In November of 1954 the provincial government responded to the townsite committee's recommendation and established the Improvement District of Manitouwadge, which encompassed an area of 230 square kilometres. Improvement district status was intended to prevent "scattered" settlement and to provide a transitional form of government for the vicinity. Taylor and Weaver of the provincial government and William Harrison of the Geco Mine were appointed trustees and charged with the responsibility for local affairs. In conjunction with the activity of the provincial government, the Board of Trustees attempted to coordinate the continued development of the comprehensive theme inherent in Manitouwadge planning. In this manner they implemented and/or continued to apply zoning regulations which recognized five different land uses: residential, commercial, service, industrial, and open space or rural.

November also witnessed the clearing and development of Neighbourhood No.1 and the completion of preliminary subdivision layouts for Neighbourhoods No.2 and No.3. In May of 1955, a "brisk" public land auction was held in Fort William, and by the summer months the community began to assume its permanent configuration.[32] As the population figures dictated, construction on Neighbourhood No.2 was begun in 1956 and on No.3 in 1957. The total townsite, which was planned for approximately 7,000 inhabitants, housed 877 persons in 1956 and 2, 635 in 1961.[33] While the townsite never actually reached its anticipated size, the involvement of the provincial government and its adherence to comprehensive planning principles generally assured the continued "model town" appearance of the community.

Elliot Lake

The community of Elliot Lake is located approximately 100 kilometres west of Sudbury. Once described as the "uranium capital of the world," the community was founded upon the uranium potential of the Algoma Basin.[34] Discovered in a contemporary sense in 1953, the mineral wealth of the vicinity gave rise at peak production to twelve uranium mines. The mines, in turn, produced what Ira Robinson has described as the "combine town" of Elliot Lake.[35] Created to accommodate the housing needs of the various mines active in the district, the townsite became the second major undertaking in community design of the two townsite committees. While it built upon the experience of Manitouwadge, it also recognized the uniqueness of the Elliot Lake mineral field. Indeed, in suggesting that the planning of Elliot Lake was "much more difficult" than the designing of Manitouwadge, the administrative subcommittee alluded to the uncertainty of the uranium market. The tentativeness was related to both the "fact that the contracts [for uranium] with the Government of Canada are for the period [only] up to 1962 and ... the uncertainty as to the role that uranium will play in the future economy of the world."[36] From its beginning, therefore, the community was very much a product of the "Atomic Age."

When between 1954 and 1956 the three major mining companies in the Elliot Lake field secured federal government contracts for the purchase of uranium, the cabinet committee committed itself to the development of a local townsite. In doing so the province rejected several other townsite development schemes, most of which provided for greater company involvement. Initial meetings between government and company officials indicated a desire, on the part of at least one mining company, to develop its own "company town."[37] The federal government also proposed the establishment of "individual company communities," but the company-town concept was quickly rejected by the cabinet committee.[38] A second planning concept, which was presented by mining magnate Joe Hirshhorn and called for the creation of the townsite under the direction of private consultants (Philip Johnson of New York and John Parking and Associates of London), was also deemed unacceptable by provincial officials.[39] In the end, the government decision on the townsite matter was largely determined by the cabinet committee's affinity for the central townsite concept. The provincial government maintained that one community would eliminate the costly duplication of services, encourage the development of a socially integrated community and "presumably would be longer-lived than the single-mine village."[40]

On 1 September 1954, provincial government authority over the development of the Elliot Lake townsite was consolidated with the creation of the Improvement District of Elliot Lake. Its 630-square-kilometre territory was intended to encompass all the scattered mine sites of the Elliot Lake field.[41] At the centre, the townsite itself covered 223 hectares (1, 619 hectares were added at a later date). Preliminary surveys, including aerial photography, soil typing and contour mapping, were conducted in October of 1954. This was done in conjunction with the activity of Algom Uranium Mines, which, according to Department of Mines records, had already conducted townsite surveys during the summer months. With the support of the Department of Planning and Development, the combined forces of the administrative subcommittee and Algom Uranium selected a townsite location "nestled between Elliot and Horn Lakes."[42]

Between 1954 and 1956 the Community Planning Branch designed and applied the "conceptual plan" of the townsite.[43] Largely dictated by topographical features, subsurface rock conditions, lines of communications and population size, the initial plan called for the creation of two townsite areas. Dubbed the eastern and western townsites, the two regions were separated by a 2.5-kilometre neck of land which was scheduled to be developed as a recreational area.[44] As it turned out, the eastern townsite – which was established as the main townsite-adequately met the community needs of the region, and plans

Map 5.1 Elliot Lake Townsite No. 1

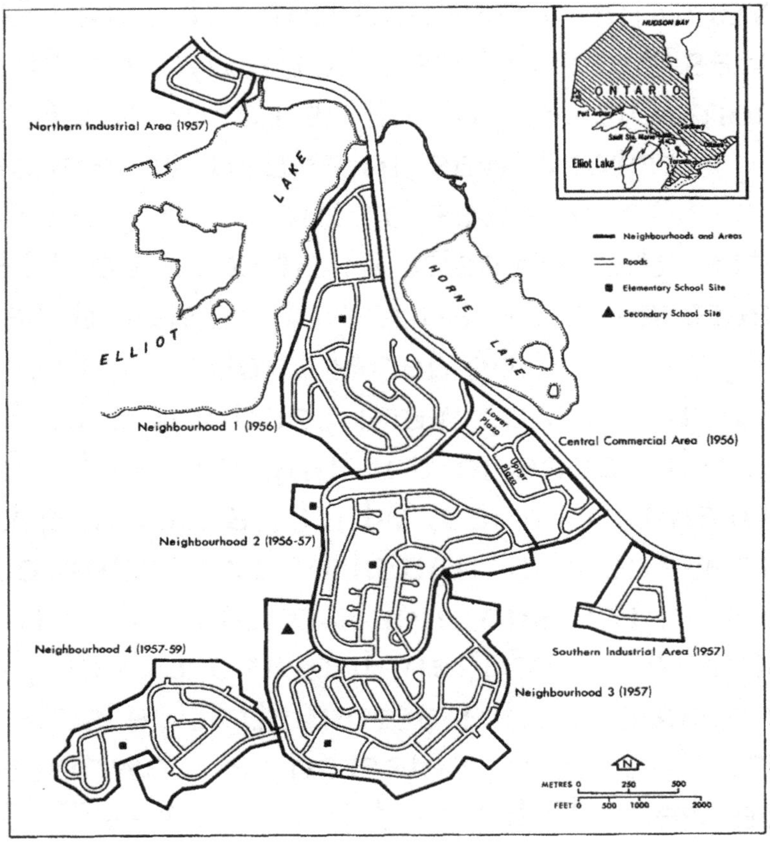

Courtesy of Léo Larivière

for the western townsite were eventually abandoned. While planning was initially undertaken for a community of 12,000-15,000, the uncertain uranium market forced the downward revision of population figures and the cancellation of the second townsite.

The townsite was designed to include three neighbourhood units, a central commercial area and an industrial zone. The configuration of the plan has been described as a crescent streetscape within the concept of neighbourhood planning units.[45] Each neighbourhood was planned to be reasonably self-contained, meeting the daily shopping, educational and recreational needs of its inhabitants. Surrounded by a predetermined greenbelt area which acted as a buffer zone between neighbourhoods, each unit stressed wide winding streets and a reasonable mix of housing design. One major thoroughfare provided access to each neighbourhood, within which secondary roadways such as avenues, crescents and cul-de-sacs served local traffic. Designed to facilitate automobile access, the overall pattern of the plan tended to draw upon suburban planning concepts, leading one critic to suggest that the community was in fact a "miserable little suburbia."[46]

The central commercial area, like the various neighbourhood units, was designed to maximize automobile access. Indeed, one source maintains that the parking area of the commercial district actually took precedence over building sites in the planning stage.[47] In this vein, property within the commercial district was first allotted to parking requirements and then "the remainder was divided into lots," almost as an afterthought, for commercial development. Rigid zoning regulations facilitated the development of the completely commercial district and through an elaborate "block system" of development, most spaces were quickly filled.[48]

The industrial area was located on a ten-hectare site that was far enough removed from the residential and commercial zones to eliminate the problems associated with conflicting zones. The area was developed adjacent to the central highway to facilitate easy access and contained approximately thirty-six lots for small industry.[49] The mines were all located some distance from the townsite and therefore presented little in the way of zoning complications.

Construction of the townsite was begun in 1956, and three years later the eastern townsite as conceived in the master plan was virtually complete. While comprehensive planning principles were generally adhered to by the provincial government, its seemingly half-hearted attention to community affairs discouraged the full development of Elliot Lake's potential. Because of its concern about the longevity of the project, the province refused to commit itself fully to the Elliot Lake townsite. Perhaps the most telling example of this can be found in the minutes of an administrative subcommittee meeting, when the president of Algom Uranium inquired as to "why the procedure followed in Manitouwadge where the government extended some $600,000 in credit... should not apply to Elliot Lake." He was informed that Manitouwadge was an "experiment" and did not constitute government policy.[50] In other words, Elliot Lake did not necessitate or perhaps warrant large-scale investment on the part of the provincial government. The so-called failures of the townsite as articulated by Norman Pearson, Rudolf Helling and others must be viewed in light of the provincial government's concern over the viability of uranium production.[51]

Cardiff

The townsite of Cardiff is located approximately ten kilometres north of the community of Bancroft. Like Elliot Lake, it was dependent upon uranium mining for its livelihood. By the summer of 1955, three uranium mines, with a total anticipated workforce of approximately 1,000, were producing ore: Bicroft Uranium Mines Limited, Faraday Mines Limited and Canadian Dyno Uranium Mines. That fall the provincial government accepted the responsibility for townsite creation, and Cardiff, like Elliot Lake and Manitouwadge, became a government-established resource townsite. Cardiff, however, offers evidence of a growing disenchantment on the part of the provincial government with its role as townsite developer.

In July of 1955, R.A. Bryce, president of Bicroft Uranium Mines Limited, approached the Department of Planning and Development with the request that the government consider the establishment of a townsite in order to "accommodate the working force, families and secondary service population which would be supported by Bicroft's mining operations."[52] Bryce further indicated that the townsite would eventually service the several mining operations in the general vicinity and that in his opinion, a centrally located townsite would allow for controlled urban development. On the basis of this inquiry the proposal was submitted to the cabinet committee for consideration in the fall of 1955. The cabinet committee in turn sought the recommendations of its administrative subcommittee. In charging the subcommittee with the responsibility for investigating the Cardiff townsite situation, the cabinet committee underscored the necessity of resolving the townsite issue. By alluding to the immediate need for residential, commercial and industrial accommodation and also by recognizing the potential land-use conflict that might develop between urban- and rural-based populations, the cabinet committee once again became a willing participant in townsite affairs.[53]

The investigative team of the administrative subcommittee ventured into the Paudash Lake area of the uranium field in October of 1955. Led by R.N. Percival, the group conducted the necessary preliminary tests and eventually determined the most logical site for a town. Located approximately 1.5 kilometres north of Paudash Lake, the site was deemed to be conveniently located for both the Bicroft Mine and a second mine site owned by Faraday Mines. Indeed Bryce of the Bicroft operation and A.W. Johnson of Faraday both informed the subcommittee that the site selected for townsite development met their approval. Its only apparent drawback was a minor drainage problem which, according to preliminary surveys, would not present an obstacle to development.

When word of the proposed townsite reached the Bancroft council, immediate counter-proposals were made extolling the virtues of Bancroft as an alternative site.[54] The ensuing debate eventually led to a sub-committee-coordinated joint meeting between mining personnel and the Bancroft council, which convened in January of 1956. The principal points arising from the meeting were that the mining companies were unwilling to consider the Bancroft alternative and that the village council saw it as the only solution.[55] In the end, the mining companies won the day and the Paudash Lake townsite proposal became the government-supported option.

The site chosen alleviated the provincial government from having to assemble land. As most of the surface rights in the townsite area were already owned by the two mining companies, the cabinet committee decided to allow private ownership of the total townsite. Bicroft and Faraday purchased the approximately 400 hectares necessary for townsite development at a cost of $30 per acre.[56] By the end of January 1956, the entire site was under the joint ownership of the two principal mining companies and construction was expected to commence in the early spring.

A report filed with the cabinet committee in May of 1956 forced a drastic revision of the Cardiff townsite scheme. Citing a study prepared by the consulting firm of R.K. Kilborn, the Faraday Mining Company withdrew its support for the proposed townsite project. In outlining both the advantages and disadvantages of the proposed location, the consultants maintained that the few advantages were "considerably outweighed by the disadvantages."[57] Perhaps more important in Faraday's decision to withdraw from the townsite scheme was the federal government tax program. After making inquiries with the Ministry of National Revenue at Ottawa, Faraday officials were informed that the cost of housing could be written off at the rate of 30 percent per annum "when located on mine property."[58] Conversely, when housing was erected on a townsite which did not form part of the mining company's property, such write-offs were not applicable. This would have been the situation with the proposed townsite because the bulk of the property was owned by the Bicroft Mining Company. To apply the scheme of tax write-offs to its fullest, the Faraday Company decided to locate its housing north of, and immediately adjacent to, the village of Bancroft on property which would allow it to exercise the tax claim.

The Faraday decision to develop a separate housing project forced a re-evaluation not just of the Cardiff townsite proposal but of government policy concerning resource-community development in general. In maintaining that the Kilborn report was "based on a very cursory examination and entirely without benefit of the extensive survey data," the administrative subcommittee reasserted government priorities in the form of a rebuttal.[59] In a report entitled "Reply to Comments of R.K. Kilborn," R.N. Percival clearly defined the position of the provincial government. Percival's report in part indicated that:

1) Government policy is to concentrate, as far as is possible, urban development necessary to serve several mining operations in one central location – reasons:

    a) Economic

    b) Adequate shopping, business and entertainment facilities are more easily attracted and developed

    c) Convenience of school children and of general family life

    d) Prevention of spoliation of countryside by ribbon and sporadic development

    e) Minimize damage to the agricultural economy of the area

    f) Problems of power, water supply and drainage can be easily overcome

2) Site chosen is as central as possible to the proposed mining activity.
3) Other available sites are not large enough.
4) Policy is to avoid straggling a main highway with urban development.[60]

In particular reference to the Cardiff townsite, Percival maintained that "this is not the most desirable piece of land for urban development in Ontario; but taking *all* factors into consideration in this particular case, no suitable alternative is available."[61]

In any case, Faraday proceeded with its plan for the erection of housing adjacent to the village of Bancroft. Dyno constructed primarily on-site facilities while also allowing its employees to seek accommodation elsewhere, and Bicroft continued with the development of the townsite. As a result of the Faraday decision the cabinet committee decided "not to reorganize municipal boundaries in the Paudash Lake area."[62] According to D.F. Taylor, plans for "one designed community" in the Cardiff area had been completely scuttled. Maintaining that the region's existing community infrastructure would adequately meet the needs of the mining industry, the cabinet committee withdrew from active participation in the townsite conundrum.[63] In general, and as alluded to by Taylor and Percival, government policy on the issue of resource-town development was returning to more of a supervisory capacity.

CONCLUSION

The Cardiff experience brought to a close the major phase of government townsite activity. Although both the cabinet committee and its administrative subcommittee remained as functional bodies, their involvement in townsite affairs dramatically declined after the Manitouwadge, Elliot Lake and Cardiff experiences. Indeed, as suggested by A.L.S. Nash, director of the Community Planning Branch in 1961: "Since the completion of Manitouwadge, Elliot Lake and Bicroft townsites there have been no large projects, but there are periodic questions ... These have been mostly dealt with by the individual Departments or Branches."[64]

The declining role of the provincial government in resource-town development was evident as early as 1956, when for a variety of reasons the cabinet committee refused to commit itself carte blanche to the development of Elliot Lake. The return to a laissez-faire townsite policy was further witnessed in Cardiff where the provincial government facilitated private ownership of the townsite property. The consequence of this was the muddled townsite situation which, perhaps more than anything, fostered government resolve to refrain from townsite development. By 1959, with the conclusion of the first stage in the Elliot Lake plan, the provincial government had returned to an advisory role in the matter of resource-town construction.

Although at various times during the tenure of the Cabinet Committee on Townsites the provincial government gave some evidence of attempting to design a resource-town policy, it was never fully successful in doing so. The various strategies pursued ranged from the laissez-faire approach of the Coldstream Copper Mine's Burchell Lake townsite right through to the complete orchestration of the Manitouwadge townsite. In no case, however, was a comprehensive policy statement issued by the provincial government. Certainly government officials recognized the benefit of a socially integrated community and the stability of a diversified local economy, but they failed to introduce the broad policy initiatives necessary to the achievement of those goals.

Notes

[1] Ontario Archives (hereafter OA), Minister of Municipal Affairs Collection, RG19, Minister's Office, Series A-3-B, Box 2, "Joint Meeting – Townsites."

[2] For a partial list of the more recent literature, see Robert Robson, *Canadian Single Industry Communities: A Literature Review and Annotated Bibliography* (Sackville, New Brunswick: Published by the Rural and Small Town Research Studies Programme, Department of Geography, Mount Allison University, December 1986).

[3] OA, op.cit.

[4] Ibid., Community Planning Branch, Series D-l, Box 170, "Towns, New – General."

[5] Ibid., Box 172, "Townsites – Mining Townsites."

[6] D.F. Taylor, "New Communities in Northern Ontario," *Housing News Letter* 3, no.2 (April 1957): 2.

[7] OA, Ministry of Municipal Affairs Collection, op. cit, Community Planning Branch – General Files, Series D-l, Box 147, "Single Industry Towns – General."

[8] Ibid., Box 170, "Townsites – Administrative Subcommittee."

[9] Ibid., Townsites, Series D-l-B, Box 20, "Moose Mountain Mine – District of Sudbury."

[10] *The Development of New Communities in Ontario* (Proceedings of a seminar sponsored by the Bureau of Municipal Research, May 1974), 12.

[11] OA, Ministry of Municipal Affairs Collection, op. cit., Minister's Office, Series A-3-B, Box 2, "Joint Meetings – Townsites."

[12] Ibid., Box 4, "Townsites – Committee."

[13] Ibid., Community Planning Branch – General Files, Series D-I, Box 170, "Townsites – Mining Townsite Kashabowie."

[14] Ibid., Townsites, Series D-I-B, Box 13, "Burchell Lake Townsite Potential."

[15] OA, Department of Mines Collection, RG 13, Minister's Files, Series A-6-A, Box 22, File 22-5, "Mining Townsites 1957-1959."

[16] OA, see Faraday Townsite, Ministry of Municipal Affairs Collection, op. cit., Townsites, Series D-I-B, Box 11, "Cardiff Townsite #2" and Pronto Townsite in ibid., Box 2, "Algom."

[17] OA, Department of Mines Collection, op. cit.

[18] OA, Ministry of Municipal Affairs Collection, op. cit., Community Planning Branch – General Files, Series D-I, Box 147, "Single Industry Towns – General."

[19] Ibid., Townsites, Series D-I-B, Box 14, "Reports on Temagami Townsite Mining."

[20] OA, Department of Mines Collection, op. cit., Series A-4, Box 12, File 12-7, "Townsite Temagami – 1956, 1957."

[21] OA, Ministry of Municipal Affairs Collection, op. cit., Minister's Office, Series A-4-2, Box 22, "Timiskaming District: Temagami Townsite."

[22] Ibid.

[23] "Manitouwadge: A New Ontario Mining Community," *Ontario Planning Supplement* 4, no.2 (1957): 3.

[24] "Here's a Unique Experiment in Modern Community Living," *Financial Post,* 18 February 1956, 59.

[25] OA, Ministry of Municipal Affairs Collection, op. cit., Townsites, Series D-I-B, Box 15, "Manitouwadge Townsite #1."

[26] Ibid.

[27] Ibid.

[28] Ibid.

[29] Ibid., "Manitouwadge Townsite #2."

[30] "Here's a Unique Experiment in Modern Community Living," op. cit.

[31] OA, Ministry of Municipal Affairs Collection, op. cit., Townsites, Series D-I-B, Box 15, "Manitouwadge Townsite #2."

[32] "Here's a Unique Experiment in Community Living," op. cit.

[33] "Population of Census Subdivisions, 1921-1971 ,"*Census of Canada* (1971).

[34] Frank Kaplan, "World's Uranium Capital Gets Hopes Badly Shattered," *Financial Post,* 20 February 1960, 3.

[35] Ira M. Robinson, *New Industrial Towns on Canada's Resource Frontier* (Chicago: Department of Geography, University of Chicago, 1962), 7.
[36] OA, Department of Mines Collection, op. cit., Series A-4, Box 12, File 12-5, "Townsite Elliot Lake – Planning and Development 1955."
[37] Rudolf Helling, *Elliot Lake Ontario* (Assumption University of Windsor, 1960), 5.
[38] R.G. Bucksar, "Elliot Lake," *Habitat* 8, no.1 (January–February 1965): 21.
[39] N. Pearson, "Elliot Lake: The Best-Planned Mining Town," *The Canadian Architect* 3, no.11 (November 1958): 55.
[40] R.G. Bucksar, "Elliot Lake, Ontario: Problems of a Modem Boomtown," *Journal of Geography* 61 (March 1962): 119.
[41] OA, Ministry of Municipal Affairs Collection, op. cit., Community Planning Branch – General Files, Series D-l, Box 172, "Townsites – Mining Townsites."
[42] OA, Department of Mines Collection, op. cit., Series A-4, Box 12, File 12-5, "Townsite Elliot Lake – Planning and Development 1955."
[43] Oiva W. Saarinen, "Single-Sector Communities in Northern Ontario: The Creation and Planning of Dependent Towns," in *Power and Place: Canadian Urban Development in the North American Context,* edited by Gilbert A. Stelter

and Alan F.J. Artibise (Vancouver: University of British Columbia, 1986), 249.

[44] H. Sears, "Report on Elliot Lake," *Journal of the Royal Architectural Institute of Canada* 35 (October 1958): 391.

[45] Rudolf Helling, op. cit., 6.

[46] Anthony Adamson as cited in O. Saarinen, "Single-Sector Communities," op. cit., 249.

[47] H. Sears, op. cit., 392.

[48] This refers to an elaborate scheme whereby the person "purchasing the first lot in a block was given the option of purchasing the remainder of the block at the same price per lot." For details see H. Sears, op. cit., 392.

[49] Ira Robinson, op. cit., 392.

[50] OA, Ministry of Municipal Affairs Collection, op. cit., Townsites, Series D-I-B, Box 3, "Elliott Lake – Algom #4."

[51] See, for example, N. Pearson, "Elliot Lake: A New Town in a Uranium-Ore District in Canada," *International Federation for Housing and Planning News Sheet,* May 1960; R. Helling, op. cit.; R.W. Jones, "Elliot Lake: A Unique Case of Community Organization," *International Review of Community Development,* no.2 (1958); and O. Saarinen, "Single-Sector Communities," op. cit.

[52] OA, Ministry of Municipal Affairs Collection, op. cit., Townsites, Series D-I-B, Box 11, "Cardiff Townsite #1."

[53] Ibid.
[54] Ibid.
[55] Ibid.
[56] OA, Ministry of Municipal Affairs Collection, op. cit., Minister's Office, Series A-3-B, Box 2, "Joint Meetings – Townsites."
[57] Ibid., Townsites, Series D-I-B, Box 11, "Cardiff Townsite #2."
[58] Ibid.
[59] Ibid.
[60] Ibid.
[61] Ibid.
[62] Ibid.
[63] Ibid.
[64] Ibid., Community Planning Branch – General Files, Series D-I, Box 171, "Townsites – Mining Townsites Kashabowie."

Chapter 6

Breakthroughs in Health and Safety in Northern Ontario Mines

Donald Trotter

Several commissions of inquiry into the health and safety of mine workers have been carried out in the Ontario mining industry. These have resulted in a probing look at how the industry has perceived its role in protecting workers. This paper touches on four of these commissions, describes why they were called, what their major findings were, and explains how the results of the inquiries have improved health and safety in the mining industry.

All four commissions were called because of an unusually high number of fatal accidents in the Ontario mining industry at certain periods of time. The following table shows the name of the inquiry, the year called, the number of fatal accidents that triggered the inquiry and the number of recommendations proposed by the inquiry.

Name of Inquiry	Year Called	Fatalities	Recommendations
Godson Commission	1928	39	15

Name of Inquiry	Year Called	Fatalities	Recommendations
Ham Commission	1976	20	117
Burkett Commission	1980	19	83
Stevenson Commission	1984	16	60

THE GODSON COMMISSION AND THE FORMATION OF THE ONTARIO MINE RESCUE ORGANIZATION

On the morning of 10 February 1928, a fire started in a stope on the 550-foot level of the Hollinger Consolidated Gold Mine. Mining had been completed in this particular stope and workers were in the habit of dumping garbage into the mined-out workings. This waste material consisted of paraffin wrapping paper, powder boxes, box lids, sawdust, fuse ends and some detonators. The fire probably started from spontaneous combustion and eventually caused the death of thirty-nine miners. All died as a result of carbon-monoxide poisoning.

At that time Ontario had no experience in fighting large underground mine fires. The special equipment required was not available, and even if it had been procured, miners had not been trained in its use. Mine-rescue assistance had to be requested from the United States Bureau of Mines in Pittsburgh. The bureau responded immediately by sending a railroad car fully equipped with special rescue personnel and breathing apparatus. Since the mine was some 1, 600 kilometres from Pittsburgh, eleven hours were required to travel to the scene. These special rescue teams were able to put out the fire, restore normal air currents and permit the safe examination of the mine and the recovery of the bodies. It was felt, however, that the unfortunate delay in reacting to the emergency caused the needless loss of many lives.

The provincial government established the Godson Commission to look at ways of preventing similar tragedies. Its fifteen recommendations are well documented and have served as guidelines for industry and government since the report was published in 1928. As a result of its major recommendation, the Ontario Mine Rescue Organization was set up.

Today this organization is recognized as an example for establishing similar programs in other provinces and countries (Smith 1988). It has a history of leadership in the development of safe mine-rescue practices. All mines now have thoroughly trained teams able to respond to an emergency by following well-established practices. Any proposed changes in mine-rescue techniques must be carefully evaluated to ensure that they are improvements over current practice. Team members are hand-picked from mine workers for their skills, physical ability and willingness to accept the strict discipline required during emergency operations. These mine emergencies could include any combination of the following: fires, explosions (methane or explosives), cave-ins, rock bursts, gas emissions, flooding, bulkhead failures, or shaft accidents. A partial list of the skills required to handle these situations includes: fire-fighting while wearing self-contained breathing apparatus, gas analysis, timbering, roof bolting, first aid, mechanical repairs, ventilation measurements and shaft repair.

The aim of the on-site mine-rescue teams is to establish an ability to deal with any emergency in its mine within four hours. In most instances the emergency is controlled in this time, but if it is not, other companies respond. All companies are willing to respond to another's distress call in times of emergency.

Funding for the mine-rescue organization is provided by the industry in two ways: A special assessment is levied by the Workers' Compensation Board (WCB) against the mining companies. This assessment covers rescue officers' salaries, building maintenance, supplies, vehicles and equipment. In addition, all costs for the individual team members, such as training, salaries and mine-rescue bonuses, are paid for by the individual mining companies. This financing is a requirement of the mining regulations.

At present, nine mine-rescue officers appointed by the Ministry of Labour are responsible for the maintenance of mine-rescue equipment and the training of teams in eight districts across the province. A senior mine-rescue officer is responsible for supervising and coordinating the mine-rescue organization. Some 1,000 miners and supervisory personnel are trained or are undergoing training in mine-rescue work. Each candidate undergoes a special mine-rescue medical examination before becoming a team member. To qualify as a team member, the candidate must complete first a basic and then a standard course. Additional courses provide advanced training for team leaders. Supervisory and management courses are also available.

Since the formation of the Ontario Mine Rescue Organization some sixty years ago, only one miner has died as a result of a mine fire.

THE HAM COMMISSION AND THE ESTABLISHMENT OF OCCUPATIONAL HEALTH AND SAFETY COMMITTEES

In 1968, unions, the public and the press noted the poor safety record of the mining industry compared with other resource sectors. The industry attempted to explain the phenomenon as being due to the inherently dangerous nature of mining. The industry also pointed out that in 1968 a change was made in the method of reporting lost-time accidents. Before that time, an accident was reported as such if it resulted in three days of lost work. In 1968, an accident had to result in only one day's lost work before it was reported. As a result of this new procedure, lost-time frequencies increased dramatically.

In 1974, the provincial government established the Ham Commission to look at all aspects of mine safety for it did not consider the industry's explanations adequate, noting an especially high accident rate among young, unskilled and inexperienced workers. Ham, a professor of electrical engineering who in 1983 became president of the University of Toronto, recommended that management be required to provide intensive training for its employees. To quote Ham:

> The apparently common view that the great majority of accidents are the direct result of nothing more than unsafe acts or unsafe conditions is, in the Commission's opinion, too restricted a view of the human problem of accidental injuries. Workmen and their supervisors at every level may act unwisely, but they do so within a system for the performance of work whose responsibility it is to set clear and supervised standards of what is expected. Within such a system workers have a clear obligation to perform work by standard procedures, and supervision has the responsibility to see that standard conditions of work, tools, and equipment are maintained.

The commission proposed that the senior management of each mining operation establish a system of accountability which clearly identified who was responsible for detecting and reporting departures from standard conditions at every level of operations; for ensuring that identified departures were dealt with; for committing resources necessary to correct anomalies; and for reviewing all actions taken.

The commission also proposed that safety auditors be drawn from the ranks of employees and a well-defined role be established for joint labour-management health and safety committees. Such committees are now required by the provincial Occupational Health and Safety Act.

External audits keep the mines' safety systems alert and responsive, but they cannot substitute for basic internal integrity which rests on the knowledge, training, experience and commitment of managers, supervisors and workers themselves. All parties are now under an obligation to seek to eliminate unsafe acts and conditions. This concept, known as the "internal responsibility system," is now well established in Ontario's mines.

THE BURKETT COMMISSION AND THE IMPROVEMENT OF THE ANALYSIS OF MINE ACCIDENT REPORTS

In 1978, the Ontario mining industry recorded only eight fatalities. The following year only five fatalities occurred. It was felt that a breakthrough in safety had been achieved, considering that the average for the industry for the previous ten years had been about twenty per year. By the end of July 1980 a complete reversal had taken place. The total number of fatal accidents for the first seven months of the year stood at eighteen, and the Ontario mining industry had to defend itself against severe criticism, particularly by the press. As a result of such publicity, the provincial government formed the Burkett Commission, which travelled throughout the province, hired consultants, talked to management, union representatives, workers and safety professionals, and in May 1981 published a two-volume report of its findings.

Burkett pointed out that while there was an abundance of opinion about accident causes, there appeared to be a shortage of facts.

At that time, provincial legislation required that a company send a report of a medical injury sustained by a worker to the Workers' Compensation Board within three days of learning of the injury. Commonly known as "form 7," the main purpose of the report was to ensure that the injured worker could receive his or her rightful compensation as soon as possible. The form was also useful for compiling statistics regarding injuries. When a worker was injured, the company sent a completed form 7 to the WCB. If the injury qualified for compensation, the WCB sent a copy of the form to the appropriate safety association. The Mines Accident Prevention Association of Ontario (MAPAO) received about 3,000 forms each year but had made few conclusions or recommendations based on its data gathering. In short, it had done little to research the causes of accidents, although with its interlocking structure of standing committees, professional staff and district safety groups, it was in a unique position to make a contribution in identifying accident causes, in addition to its more traditional role of education and consultation. To quote from Burkett, who put it bluntly:

> The Association, although collecting and publishing accident statistics, does little follow-up and analytical work in order to identify the factors underlying accident trends ... The MAPAO should play a role in identifying factors which may be inhibiting safety performance, and must be prepared to lay the ground work for reassessment and re-evaluation where appropriate.

Burkett was not alone in his criticism of the mine accident reporting system. In 1978, the Gibbs Report had come to essentially the same conclusion. Commissioned to describe what was then known about health and safety in the Canadian mining industry, Graham Gibbs severely criticized the reporting of accident data:

> As the standardization of reporting methods and the analysis of data for publication involve considerable work, it is essential that any attempt at producing data in the provinces or nationally

be coordinated with persons most likely to use them, such as the mining industry, researchers, and government agencies responsible for mining. Much valuable time and effort can be spent on producing data of little value to anyone. This has been a common problem in reporting accident statistics. Indeed, the problem exists for all industry sectors in Canada, and the lack of published data has been an object of frustration to some government departments in obtaining estimates of the extent of the industrial accident problem.

And it was slowly dawning on the mining industry itself that the existing procedures for the reporting of accident data were unsatisfactory.

In July 1980 at a meeting in Sudbury sponsored by the Ontario Mining Association, eighty-eight senior mining industry representatives thoroughly reviewed the circumstances surrounding the high number of fatalities. They agreed to fourteen specific recommendations, one being that there should be a systematic audit of company safety programs. As a result, the MAP AO arranged through the Industrial Accident Prevention Association to be licensed to conduct audits, a program which had been well tested in South Africa and subsequently used in other countries. The MAPAO began conducting audits in 1981 after completing the necessary licensing arrangements. Two of the components of the audit program were accident/incident investigation and accident/incident analysis. Average scores on these components were low, substantiating the need for a standard investigation reporting system.

In late 1981, as a follow-up to the Burkett Report, the MAPAO sent out a questionnaire concerning the accident report forms being used in the industry. The purpose of the questionnaire was to determine to what extent accidents were being investigated, to what extent the industry was trained in investigating accidents, and whether it was possible to have a common accident investigation form.

As a result of the replies received from the questionnaire, the MAPAO decided to design a standard accident report form for the use of the mining industry. This determination was given added impetus a couple of years later by the Weiler Report to the Ministry of Labour which identified the need for more comprehensive injury reporting. In recommendation 2.5, Weiler states:

> During the course of my inquiry, I have strongly recommended to the WCB, the Ministry of Labour, and the nine provincial safety associations that they work together to develop a better form for employer reporting of workplace injuries. Ideally this should be a two-part form, the first part of which would contain the bare information needed to process the compensation claim, while the second part – to be completed and filled out later by the employer – would contain more elaborate details of the circumstances of the in-

jury, which would help in understanding, and then controlling, the important recurring causes.

At this writing, the form has been in place for six years and is proving a valuable tool in getting at the underlying cause of accidents (see also Cluff 1983; Trotter 1984). It is used for every medical injury in the mining industry, requiring the computer-coding of some 7,000 forms each year. The accident investigation form has proved to be a good investigation tool from which information can be collected, stored, sorted and cross-referenced by computer. This has helped prevent recurring accidents. Medical-injury frequency has declined dramatically. The very act of carrying out a thorough investigation of every medical injury has helped instill a safety consciousness in those involved and in itself has led to a reduction in accident frequency.

TRAINING AND COMMUNICATIONS

The Ham Commission in 1976 and the Burkett Commission in 1980 emphasized training and communications as key elements in the fight to prevent accidents.

In 1983, the Mining Health and Safety Branch of the Ontario Ministry of Labour undertook a program to meet these training and communication needs (Pakalnis 1983). The primary target groups were identified as: joint health and safety committee members, Mining Health and Safety Branch inspectors, mining industry supervisors, worker inspectors, and

MAPAO area representatives – in total about 4,000 persons.

Through these select groups it was possible to reach the 50,000 or so workers in the Ontario mining industry. Thirty training modules were agreed upon, based on the 283 sections of the regulations which contain minimum performance objectives in the many complex operations involved in mining. These sections were not always completely understood by all parties concerned and inconsistencies had arisen in the application of various regulations to certain situations.

In order to determine the content for each of the thirty modules, tripartite groups were established, composed of managers, workers and government officials. Each module covered specific subject matter and included the relevant regulations, appropriate reference material, necessary theory and practical material, and clear learning objectives. The key objectives of the training modules were to teach pertinent regulations, performance objectives, safe practices, proper terminology, related hazards, and inspection competence.

This program was ambitious because it included subject areas in which consensus was sometimes elusive and technical solutions were sometimes undeveloped. It was a practical program, aimed at real people using approaches that reflect existing levels of expertise. It provided a neutral corner in which government, industry and labour can discuss and agree on what is required by the mining legislation. Further, it provided a comprehensive look at the subject area, identifying hazards clearly and illustrating in slides or graphics acceptable practices for Ontario mines.

THE STEVENSON COMMISSION

A rock burst on 20 June 1984 at the Falconbridge Mine near Sudbury caused the failure of the roof in an undercut-and-fill stope killing four miners. Rock bursts are caused by the failure of walls, pillars or the mine roof, and result in a sudden, explosive release of energy. This particular rock burst ranged in magnitude from 3.0 to 3.5 on the Richter scale. Because of the severity of the burst the mine was closed and 400 workers were laid off.

The next month on 6 July 1984, a rock burst of 4.0 on the Richter scale took place in Inco's Creighton Mine. Fortunately, the mine was shut down for the summer vacation and there were no casualties. A series of rock bursts also occurred on 17 December in the Quirke Mine in Elliot Lake. Similar rock bursts were reported from other mines in the Red Lake area and in Kirkland Lake.

As a result of these incidents, in 1985 the provincial government established the Stevenson Commission to look into emergency preparedness and ground control, that is, into the methods that are used to ensure that the underground is safe and stable. Its report was the result of seventeen months of work which included forty-six public hearings, visits to fifteen mines, and consultations with mining companies, local and international unions, government agencies, mining trade associations and mining research groups.

Many of Stevenson's key recommendations reflected the concerns of earlier commissioners and included a call for the improved training for miners, supervisors, worker inspectors, and non-miners working underground; improved communications among all parties in the workplace for both routine and emergency situations and improved communications equipment; expanded mine-rescue training; and equipment for other than fire emergencies. But Stevenson also called for improved rock mechanics programs at Ontario colleges and universities and for the establishment of a research institute in ground control to coordinate ongoing and future research.

Many of the recommendations have now been implemented, including the establishment of the Geomechanics Research Centre at Laurentian University. Work is progressing on several fronts to avoid the sudden and extremely dangerous stress-release problems that mines have experienced in the past.

Management is now acutely aware of the hazards of poor long-range planning resulting in dangerous stress build-up over a long period of time. Although no one is claiming the stress problem in mines is solved, the problem is certainly recognized, understood and being tackled on several fronts. As knowledge increases and corrective steps are taken, the chances of further tragedies lessen.

CONCLUSION

The struggle to reduce industrial accidents is a never-ending one. Justice Godson, in his closing remarks, stated: "The lessons we learn in times of sorrow and trouble are well learnt, and last for a great while, even though the memory of mankind be notoriously short." Past experience shows that attention to health and safety matters is triggered by tragedy. This is regrettable and one wishes the industry had the foresight to take necessary corrective actions before it was jolted into reaction. The actions taken, however, are giant steps in the right direction. Ontario's mining industry presently has an enviable safety record.

References

Burkett, Kevin.1981. *Towards Safe Production: The Report of the Joint Federal-Provincial Inquiry Commission into Safety in Mines and Mining Plants in Ontario.* Toronto: s.n.

Cluff, James.1983. *The Development of a Province-Wide Accident Investigation Reporting System.* Proceedings of the Fifty-second Annual

Technical Sessions, Mines Accident Prevention Association Ontario. North Bay: MAPAO.

Gibbs, Graham.1978. *Health and Safety in the Canadian Mining Industry.* Kingston: Queen's University, Centre for Resource Studies.

Godson, T.E.1928. *Royal Commission of Inquiry into the Hollinger Mine Fire.* Toronto: The Commission.

Ham, James.1976. *Report of the Royal Commission on Health and Safety of Workers in Mines.* Toronto: Ministry of the Attorney General.

Pakalnis, Victor.1983. *The Ontario Mining Health and Safety Training Program.* Proceedings of the Fifty-second Annual Technical Sessions, Mines Accident Prevention Association Ontario. North Bay: MAPAO.

Smith, Malcolm.1988. *Ontario Mine Rescue – Are You Prepared?* Proceedings of the Fifty-seventh Annual Technical Sessions, Mines Accident Prevention Association Ontario. North Bay: MAPAO.

Stevenson, Trevor.1986. *Improving Ground Stability and Mine Rescue: The Report of the Joint Federal-Provincial Inquiry Commission into Safety in Mines and Mining Plants in Ontario.* Toronto: Queen's Printer.

Trotter, Donald.1984. *Accident Investigation in the Ontario Mining Industry.* Proceedings of the Annual General Meeting of the Canadian Institute of Mining and Metallurgy. Montreal: CIMM.

Weiler, Paul. 1983. *Protecting the Worker from Disability: Challenges for the Eighties.* Toronto: Ontario Ministry of Labour.

Chapter 7

Government Regulation of Health Hazards in the Ontario Uranium Mining Industry, 1955-1976

Brian Walker

The increasing intervention by the state in the economies of many advanced industrialized countries in the post-World War II period brought about a considerable growth in the number of independent state boards and agencies authorized to regulate industry. The establishment of these boards and agencies appeared to give economic regulation greater independence from the political process and the influence of industry than had hitherto been the case. However, the creation of such agencies and boards also resulted, in some cases, in a strange irony. Their policies came to reflect closely the interests of the industry they sought to regulate rather than the "public" interest. A case in point was the Atomic Energy Control Board (AECB) of Canada's regulation of health hazards in the Ontario uranium mining industry. The policies pursued by this agency of the Canadian state were in many ways harmful to the health of mine workers in the industry.

This paper examines those factors which can account for the state's inadequate regulation of radiation hazards in the Elliot Lake uranium mines. The main factors identified are the Canadian state's interest in uranium sales, the potential commercial value of the Canadian-designed and-built CANDU reactor, and the organizational structure of the Ontario Department of Mines. In addition, the Canadian nuclear industry was a closed, tight-knit community, in which AECB members identified closely with state-owned nuclear firms. Moreover, the regulatory process and the research priorities of the AECB resulted in it being dependent on the uranium mining industry for the development of control technology.

Before analyzing in greater detail those factors which influenced the Canadian and Ontario governments' regulation of health hazards in the Elliot Lake uranium mines, let us briefly review the history of the state's efforts to regulate health hazards in these mines

CHRONOLOGY OF EVENTS

From as early as 1958 – only three years after the opening of the first uranium mine in the Elliot Lake area – union officials were aware of the potential health hazards to which uranium mine workers were being exposed, and consequently sought to persuade the Ontario Department of Mines, the Ontario Department of Health, and the Canadian Department of Health and Welfare to undertake measures which would reduce workers' exposure to radiation and silica dust (United Steelworkers of America [USWA] 1973, 1974, 1976). Until 1974, however, there was little protest from rank-and-file members about these hazards, chiefly because of economic uncertainty about the future of the uranium mines (Tataryn 1979).

What were the uranium mining companies able to do to create healthier workplace conditions? Average radiation levels remained above the mining industry's own guideline until 1967, twelve years after the opening of the first uranium mine (Ontario, Royal Commission 1976). If we turn to the other important health hazard, silica dust, average dust levels for much of the period after 1955 were also above the industry's own guideline (Ontario, Royal Commission 1976). The mining companies were extremely slow in reducing both radiation and dust levels.

From the mid-1950s to 1974, the major regulatory agency of the Canadian atomic energy industry and of the Elliot Lake uranium mines, the AECB, did not monitor environmental conditions in these mines. Instead, it relied on the Ontario Department of Mines for such services (Ontario, Royal Commission 1976). The AECB was thus ineffective in bringing about a rapid decline in radiation levels in these mines; in fact, it was not until 1969 that average levels met the industry's own permissable levels. Similarly, the agency failed to consistently reduce dust levels below the industry's own guidelines until 1976.

The Ontario Department of Mines, for its part, did not issue guidelines for radiation exposure limits until 1967, and did so then only in response to actions in another jurisdiction (Tataryn 1979). The department never issued permissable limits for silica dust, but merely accepted those issued by the provincial mining industry.

In the early 1970s, with the beginnings of modest growth in the Elliot Lake uranium mines after years of stagnation, union leaders began to protest to the Ontario ministries of Mines and of Health and to the federal Ministry of Health and Welfare. In November 1973, union leaders representing workers at the only two companies still operating in the Elliot Lake area asked for a health survey of all 1, 500 workers employed at the two companies. In response, the Ontario Ministry of Health in February and March 1974 undertook a survey of 1,000 workers to discover the nature and extent of any respiratory diseases from which they might be suffering. The results of this survey, which were published by the Ontario government one year later, in April 1975, showed that of the 1,000 men surveyed, 3.6 percent had silicosis and 5.3 percent were suffering from a pre-silicotic condition (Ontario, Royal Commission 1976). Union leaders, however, were not satisfied with the scope of the Ontario government's survey and demanded that the federal minister of health and welfare, Marc

Lalonde, ask the Ontario Ministry of Health to undertake a survey of all workers employed at the two mines. On 16 April they also met with Leo Bernier, the Ontario minister of natural resources, to request the same thing and, in addition, stricter enforcement of health regulations at the two mines (USWA 1974).

Before either union leaders or Bernier could take any additional action, however, rank-and-file members went on an "unofficial" strike at Denison Mines Limited on 18 April and remained off the job for eighteen days. In the Ontario legislature the New Democratic Party raised the issue of high levels of dust and diesel fumes in the uranium mines. A number of NDP members, during a debate in May on the estimates of the Ontario Ministry of Natural Resources, seized the opportunity to make a concerted attack on the ministry's policies and practices on occupational health and safety in the Ontario mining industry. At that time, Leo Bernier made a commitment to the committee that he would announce on 5 June "a program of action dealing with the health of workers in mines, with particular reference to the uranium mines" (Ontario, Estimates of the Ministry of Natural Resources 1974). As he had promised, Bernier on that day announced to the legislature that a royal commission would be established to study not only the particular problems of the Elliot Lake uranium mines, but those of the province's entire mining industry. The minister assured the legislature that the commission would be empowered to recommend major changes in the legislation governing health and safety in the industry (Ontario, Legislative Assembly 1974). On 18 September 1974, William Davis, the Ontario premier, appointed Dr. James Ham, a University of Toronto professor, to head the royal commission. Hearings of the commission, which began in January 1975, were held throughout the province and were almost completed by June.

What actions did the AECB take in response to these concerns? In June 1974, it incorporated new radiological provisions into the Atomic Energy Control Regulations. These amended regulations, as well as stipulating requirements for protective procedures, instruments and equipment, specified a maximum permissable dosage for ionizing radiation (Science Council of Canada 1979). However, these provisions did not cover uranium mine workers. Only in January 1978 was the Atomic Energy Control Act amended and uranium mine workers declared to be atomic radiation workers and henceforth afforded the protection of a more stringent statute on health and safety matters. The revised regulations also specified that companies had to submit summaries of radiation, dust levels and workers' exposure to them (Science Council of Canada 1979). In November 1974, the board reviewed its procedures for the licensing of uranium mines and established a Mine Safety Advisory Committee. Its mandate was to make recommendations to the board about what health and safety conditions should be satisfied before licences for uranium mining were issued. Despite this reform, two well-known radiation experts said that difficulties in getting the compliance of the uranium mining companies and the nuclear industry still remained. The AECB's research and compliance programs remained weak, as did those of the provinces. They also criticized the composition of the Mine Safety Advisory Committee, which was entirely comprised of scientific experts and federal and provincial government officials, but included no workers (Science Council of Canada 1979).

Even after the "unofficial" strike at Denison Mines, the Ontario government was unable to bring about a rapid improvement in environmental conditions in the Elliot Lake uranium mines. Even as the hearings of the royal commission were proceeding, there was no improvement in conditions. On 13 March 1975, almost one year after the strike, the NDP told the minister of natural resources, Leo Bernier, that in a survey of dust levels carried out in January 1975, sixty-seven of the seventy-seven samples taken registered above the industry's own guideline, as did eight out of ten working areas in the mines (Ontario, Legislative Assembly 1975). On 6 June the NDP pointed out to Frank Miller, minister of health, that despite assurances from Leo Bernier that environmental conditions would quickly improve in the uranium mines, Peter McCrodan, director of the Mines Engineering and Inspection Branch of the Ministry of Natural Resources, had stated in a submission to the royal commission, on 4

June, that Denison Mines "had not yet come near meeting a target standard for dust levels" set by the same ministry (Ontario, Legislative Assembly 1975). In fact, it was not until near the end of 1976 that dust levels could be regarded as satisfactory.

THE ATOMIC ENERGY CONTROL BOARD OF CANADA

How can we account for these regulatory inadequacies? The roles of the AECB are described, in large measure, in the Atomic Energy Control Act, first enacted in 1946. The act authorizes the AECB to regulate and control atomic energy materials, equipment and information in the interests of national and international security, to control atomic energy materials and equipment in the interests of public safety, and to promote atomic energy research. Since the act was passed shortly after World War II, it showed an overwhelming concern for the security of strategic materials used in the production of nuclear energy. The act conferred on the AECB and the cabinet a wide range of powers. These included the power to regulate, to license, to revoke or suspend licences, to expropriate, to create crown corporations and to give grants for research and development. The scope and breadth of the AECB's powers are extraordinary when compared with other federal regulatory agencies (Canada, Law Reform Commission 1977). But it should be noted that the act lays particular emphasis on security rather than on the protection of workers falling within the regulatory jurisdiction of the AECB. Such an emphasis tended to inhibit the passing of statutory provisions which would have afforded greater protection to such persons.

Many of the AECB's powers are exercised through the Atomic Energy Control Regulations, which again reflect an enormous concern about security (Canada, Law Reform Commission 1977). The AECB is not permitted to disseminate information about atomic energy without ministerial approval. A concern about security is again reflected in the absence of provisions for public hearings, judicial inquiries and judicial adjudication in the board's decision-making process (Canada, Law Reform Commission 1977).

The Atomic Energy Control Board of Canada was established in 1946 by the Liberal government of Mackenzie King. In the post-World War II period the Canadian government was deeply concerned about the spread of nuclear materials and nuclear weapons. The original intent of the government was that the AECB, under the presidency of General Charles McNaughton, would exercise a moderating influence on the increasing military rivalry and mounting tension between the United States and the U.S.S.R. However, the federal government was also intent on ensuring that the infant nuclear industry and its own crown corporation, Eldorado Refining and Smelting Company, would prosper. Thus, the creation of the AECB occurred in a post-war context in which the government was pursuing two objectives which appeared to be contradictory. It was intent on encouraging the sale of uranium to the U.S.A. and Great Britain to enable them to develop their own atomic energy and nuclear weapons programs while seeking, at the same time, to limit the spread of nuclear weapons (Bothwell 1984). It would appear that the government wished to confine the development of nuclear energy and nuclear weapons to only a small number of countries which were part of the Western alliance.

In the 1950s, private operators in Ontario wanted to mine and develop uranium deposits. The AECB held discussions with officials of the Ontario Department of Mines about licensing arrangements. At this time the AECB was primarily interested in the security of uranium and information about reserves, production and location. It was understood that provincial authorities would take responsibility for the health and safety of workers in these mines. It was also agreed that the AECB, in granting its exploration and mining licences, would impose a condition requiring compliance with provincial laws on mine safety (Ontario, Royal Commission, *Hearings* 1976).

In this period the AECB encouraged the development of uranium mines. The unprecedented sale of nuclear materials to the U.S.A. and Great Britain made stringent regulation of health and safety conditions in the Elliot Lake mines between 1955 and 1959 extremely difficult because the operating companies had to meet strict deadlines for the supply of uranium materials (Doern 1978; Ontario, Royal Commission 1976).

The manner in which the AECB carried out its regulatory functions was a product of its evolution and adjustment to changing conditions, not only in the immediate post-war period, but also in the later period in which Canada sought to develop an atomic energy program for peaceful purposes. Its later evolution in the 1960s was conditioned by the economics of CANDU, the Canadian-designed and-built reactor. CANDU represented a Canadian technological development with potentially high economic rewards. It was extremely important politically as well because the Canadian government was eager to demonstrate that state-run enterprises could be economically viable. In economic terms it represented important export potential developed largely by a Canadian state enterprise, Atomic Energy of Canada Limited (AECL) (Doern 1978).

In addition to the AECB's role in the promotion of uranium sales and the development of the CANDU reactor, a second factor reinforced what could be called its role as a quasi-promoter of the nuclear energy industry. This was its performance in the development of research technology. It failed to develop research technology which would have enabled it to play an effective role in the prevention of occupational diseases in the uranium mines. Regulators need the technology that can make regulatory objectives achievable in practice. Technology is required first to make the workplace safe (for instance, through the development of superior ventilation equipment), and then to ensure regulatory compliance by developing accurate devices to measure radiation and dust levels. However, the research priorities of the AECB were, in this period, focused on basic nuclear physics. It left the development of control technology in general to the nuclear power industry and to the uranium mining companies. This contributed to an excessive dependence on the industry which it was authorized to regulate (Doern 1978).

By far, the largest part of the AECB's budget was spent on fulfilling its statutory mandate to establish grants-in-aid for research in atomic energy. These grants were targeted to pure or basic research as opposed to more applied research to support its regulatory functions. Not until about 1975 did this change. About 80 percent of the AECB's budget until then went to its basic research-oriented granting program. This policy continued to be pursued despite the fact that almost all the business time of the AECB was devoted to regulatory activities (Doern 1978).

The tendency of the AECB to be a promoter of the industry was further strengthened by the close identification of AECB members and state-owned nuclear firms. There was a close association between presidents of the AECB and officials of these firms. After General McNaughton retired in 1948, he was followed until 1975 by presidents who had been members of state-owned corporations in the nuclear industry. Prior to 1975, the presidents of AECL and Eldorado had served on the AECB for fourteen years (Canada, Law Reform Commission 1977).

A third factor which had tended to make the AECB a quasi-promoter of the nuclear industry until about 1974 was the closed nature of the nuclear community and the closed nature of the regulatory process itself. The AECB was created in the post-war period when the Canadian nuclear community was a very small group of experts composed largely of members of the National Research Council and the AECL. Although by the mid-1960s this was changing as more expertise developed within the universities, until the mid-1970s, those consulted by the AECB tended to be the same small number of experts from industry and government agencies (Doern 1978).

ONTARIO DEPARTMENT OF MINES

We have examined those factors which conditioned the AECB's regulation of the Elliot Lake uranium mines. Let us now look at the part played by the Ontario government. Although the Elliot Lake uranium mines were located in Ontario, the British North America Act allotted control over them and responsibility for health and safety to the Canadian government. However, through jurisdictional accommodations, the AECB made administrative arrangements that gave Ontario this authority. The Ontario Department of Mines and its 1972 successor, the

Ministry of Natural Resources, assumed responsibility for health and safety in the Elliot Lake uranium mines until 1976 (Ontario, Royal Commission 1976). However, a number of factors tended to diminish their regulatory effectiveness.

It should be noted that during the 1960s there were many requests from provincial mines ministers for the Canadian government to transfer responsibility for the uranium mines to provincial jurisdiction. The Canadian government, however, stated that although the Atomic Energy Control Act established federal government jurisdiction in the areas of national security and foreign policy, this should not limit provincial provisions relating to other matters in the uranium mines. Moreover, the AECB's mining permits would only be issued contingent upon licences complying with all provincial regulations pertaining to these mines. The Canadian government also stated that the AECB would appoint provincial officials as inspectors whenever possible under the health and safety sections of its regulations. Moreover, since health and safety matters were primarily the responsibility of the provinces, the AECB urged Ontario, through the Dominion Council of Health, to issue radiological safety standards. However, Ontario did not take this step (Doern 1978).

Most of the resources of the Ontario Department of Mines and its successor, the Ministry of Natural Resources, were focused on promoting and furthering the interests of the mining industry in Ontario. Only a small part of it – the Mines Engineering Branch – was devoted to auditing and monitoring mining health and safety conditions. But its expertise lay in engineering and in mine safety, not disease prevention (Ontario, Royal Commission 1976). The various reformed versions of the Ontario Mining Act, which regulated the industry until 1976, lacked any precise provisions covering health requirements. The act contained only very general provisions on health and safety, and as the Ham Commission observed, most of these defied precise legal interpretation (Ontario, Royal Commission 1976). Thus, in the final analysis, the uranium mining industry was regulated by non-statutory guidelines which had no authority in law and could not be legally enforced. Adherence to them could only be sought by the Ontario government through moral suasion. In reality, the mining industry in Ontario was self-regulated, and its management alone was responsible for health and safety.

Most of the government administrative bodies which regulated the Elliot Lake uranium mines had a variety of functions which encouraged organizational fragmentation and ultimately limited the state's regulatory effectiveness.

CONCLUSION

To explain the Canadian government's inadequate regulation of radiation hazards in the Elliot Lake uranium mines, the institutional context in which regulation took place must be examined. Regulation was shaped by the AECB, whose conception and evolution occurred in a post-World War II era. At this time, the overriding concern was with the future security of atomic energy and uranium. This was accompanied by an interest in promoting the sale of uranium and the CANDU nuclear reactor. The Canadian government's interest in uranium sales and in the international sales potential of CANDU gradually made it into a quasi-promoter of the uranium mining industry, a tendency that was reinforced by the research priorities of the AECB, the close association of the AECB's members and those of the state-owned nuclear companies, and the closed nature of the Canadian nuclear community.

The Ontario Department of Mines and its successor, the Ministry of Natural Resources, had responsibility for controlling radiation and silica dust levels in the uranium mines until 1976. They were administrative units whose main resources were devoted to the promotion and expansion of the mining industry and not to the creation of a more sanitary workplace. Moreover, the province's administrative response was conditioned by the provisions of the Mining Act, which placed responsibility for health and safety in the uranium mines on mine management. It is within such an organizational and juridical context that the regulatory inadequacies of the Atomic Energy Control Board of Canada and the Ontario Department of Mines, however inexcusable, become, at least, comprehensible.

References

Bothwell, R. 1984. Eldorado: Canada's National Uranium Company. Toronto: University of Toronto Press.

Canada. Law Reform Commission. 1977. The Atomic Energy Control Board. Ottawa: Law Reform Commission.

Doern, B. 1978. "Science and Technology in the Nuclear Regulatory Process: The Case of the Canadian Uranium Miners." *Canadian Public Administration 21,* no.1: 51-82.

Ontario. Estimates of the Ministry of Natural Resources. 1974. *Standing Committee on Supply,* S840-920. Toronto: Queen's Printer.

_____. Legislative Assembly. 1974; 1975. *Debates,* 29th Legislature, 4th and 5th Sessions. Toronto: Queen's Printer.

_____. Royal Commission on Health and Safety in Mines. 1976. *Hearings.* Toronto: Ministry of Attorney-General.

_____. Royal Commission on Health and Safety in Mines. 1976. *Report.* Toronto: Ministry of Attorney-General.

Science Council of Canada. 1979. *An Overview of the Ionizing Radiation Hazard in Canada.* Ottawa: Ministry of Supply and Services.

Sims, G. 1981. A History of the Atomic Energy Control Board. Ottawa: Ministry of Supply and Services.

Tataryn, L. 1979. *Dying for a Living: The Politics of Industrial Death.* Ottawa: Deneau and Greenberg.

United Steelworkers of America. 1976. *History of Steelworkers' Action for Health and Safety in the Ontario Mining Industry.* Toronto: USWA.

______. 1973; 1974. *Steel Labor.*

Part Four

Present Challenges

Courtesy of Town of Kirkland Lake

Mayor Joe Mavrinac

Courtesy of City of Elliot Lake

Mayor George Farkouh

Courtesy of *Temiskaming Speaker*

Reeve Ron Prefasi

Introduction

In the late 1980s, several Northern Ontario communities were staggered by mine closures or suffered severe cut-backs in their mining workforces. They responded with self-help strategies and appeals for government assistance. These recovery efforts were spearheaded by their local municipal leaders, Mayor George Farkouh of Elliot Lake, Mayor Joe Mavrinac of Kirkland Lake and Reeve Ron Prefasi of Temagami. In the following series of articles, the three present realistic yet compellingly optimistic portraits of their communities and their needs, portraits that in critical ways contrast with the more pessimistic emphasis upon impersonal trends in academic analyses of the situation.

The first paper, prepared in October 1990 by Mayor George Farkouh, describes how Elliot Lake was attempting to cope with the loss of over 2, 500 mining jobs in the uranium industry. A key component of the community's strategy was to convince Ontario Hydro to continue its long-term contract with Denison Mines. By the spring of 1991, however, it was evident that this strategy had failed. By deciding to end the contract, Hydro forced Denison to cut another 1,050 jobs early in the second quarter of 1992. Even in responding to this news, Farkouh was upbeat. "This is the second blow, a left hook, which they threw us," he is reported as saying, "Now we have to get off the mat and start fighting again" (*Globe and Mail,* Toronto, 30 April 1991, 1).

This same positive thinking is reflected in the contributions of two other Northern Ontario civic leaders. In his paper, Joe Mavrinac, the long-time mayor of Kirkland Lake, notes that down-sizing in the mining industry has characterized much of his community's history. Therefore, the more recent closure by Dofasco of the Adams iron mine in 1989, while surprising, was not entirely earth-shattering. At the time of the conference in October 1990, the highlight of Mavrinac's strategy for Kirkland Lake, of course, was his plan to turn the former Adams mine site into a storage depot for the garbage of Metro Toronto. As with Farkouh, these plans have since come undone because of external forces beyond his control. In Mavrinac's case, the new provincial NDP environment minister, Ruth Grier, supported by the Temiskaming Anti-Garbage Coalition, announced that "the Ontario government will not allow Metro Toronto's garbage to be dumped elsewhere in the province as a long-term solution to the region's garbage crisis" (*Globe and Mail,* 3 April 1991, 1).

In the third and final paper in this part, Reeve Ron Prefasi of Temagami reflects philosophically on the closure of the Sherman mine, also owned by Dofasco. In his community's case, Prefasi notes, mining has not been "down-sized," it has been eradicated. Temagami is, of course, a community whose fortunes are being affected by more than mining. What is particularly interesting about Prefasi's piece is that it shows how civic leaders planned for the closure by using mining revenues to diversify the area's future economic base. And just as interesting is the fact that, for reasons outlined in the paper, some members of the community actively opposed such spending plans.

Chapter 8

Elliot Lake

George Farkouh

The major economic disaster that Elliot Lake has faced over the past ten months is incredible. The loss of over 2, 500 mining jobs, the equivalent of over one million jobs in Metro Toronto, represents 70 percent of our current mining workforce. This disaster hit with tremendous force, and might have heralded the end in lesser communities. We, however, represent a tougher breed of people; individuals who have been hardened by the realities of the North, tempered by the cyclical nature of the mining industry, and inbred with an individualistic spirit unwilling to succumb to temporary setbacks.

The Chinese have a saying that we are "cursed to live in interesting times." If this is so, as mayor of Elliot Lake I am indeed a cursed man. I and members of council would have it no other way, however. As the leadership in our community, we are dedicated to the survival of Elliot Lake and to the continued well-being of our citizens.

In managing the down-sizing of Elliot Lake, it is important to remember the scope of the problem. Elliot Lake, with a population of over 16,000, represents the largest single-industry town in Ontario and in fact is the sixth-largest centre in Northern Ontario.

The magnitude of our problem is enormous because our situation combines both the problems of a mining town (exhaustible ore bodies, wealth dependent on fluctuating world markets, and relative isolation) with those of a medium-sized city (a substantial housing stock, second-and third-generation citizens and an enormous infrastructure).

In the case of Elliot Lake, these problems have been further exacerbated by a commitment made by Ontario Hydro in 1978 to purchase sufficient uranium from Elliot Lake to support a population base of 30,000 through the year 2020. Based on this commitment, the town undertook the construction of new homes, new roads and new schools, in fact, a complete new infrastructure built for the estimated increase of population. Doing this has left Elliot Lake with a capital debt burden of $11 million at 17 percent (which has been reduced to $8.5 million) and a shrinking tax base in danger of becoming incapable of serving the debt.

As the mayor and council in a community experiencing significant economic displacement, we have had three main priorities:

1. Maintaining the town's essential services at a cost capable of being borne by a shrinking tax base;
2. Reassuring the population by providing assistance to those forced to leave the community and

building confidence among those wishing to stay; and

3. Providing a vision of economic diversification and leadership to the community to encourage economic initiatives.

One of the first things we did was commission a consultant's report to advise council on how to reorganize the town's administration to ensure that all areas of the corporation were operating at top efficiency. We wanted to make sure we were receiving maximum value for our money. In cooperation with the senior administration of the town, we have recently completed our review of the report and are embarking on a number of reforms which will result in considerable cost savings without the need to reduce services significantly.

Secondly, we have provided the leadership and discipline necessary to ensure maximum financial prudence. We have insisted that sufficient reserves be maintained and that provisions be established for all questionable sources of revenue. Our recent audit report, tabled by Peat Marwick, was most complimentary and emphasized the financial integrity of the town.

Thirdly, we have made strenuous efforts to convince Ontario Hydro that they have a moral obligation to provide financial assistance to our community and to continue to purchase all their uranium needs from Elliot Lake. The capital debt we incurred in expanding is the direct result of the request of Ontario Hydro to develop an infrastructure within which they could conduct their mining operation.

It is critical to remember that the sacrifices made by the residents of Elliot Lake have been a key ingredient in the well-being of this province, having provided the source of inexpensive and dependable energy upon which the economic expansion of the last seven years has been largely based.

The second priority has been to reassure the population. From the moment of the first Rio Algom announcement in January 1990, we have tried to ensure that a proper balance is maintained between the economic realities of the layoffs and the real possibility of renewal in the future. It is important to ensure that no resident of Elliot Lake remains because of unrealistic promises or that no one wishing to stay leaves because of undue pessimism. We worked closely with the business community through the Chamber of Commerce, the union and other groups to maintain their confidence and to encourage them to respond to the situation as a challenge and not as a disaster.

One of my greatest challenges in this respect has been my relations with the media. For the past eight months, Elliot Lake has been the centre of attention of national, provincial and regional coverage. Our community has been the subject of stories on such national programs as "Venture," "Sunday Morning" and "CBC Newsworld." Coverage has been extensive in both the *Toronto Star* and the Toronto *Globe and*

Mail, and, of course, articles appear almost weekly in the regional press.

I have personally, and I believe necessarily, coordinated the community's approach to the press to ensure that we sing from one song sheet, with one theme, with one message. I believe we have been successful in projecting Elliot Lake not as a community in decline, but rather as one in transition.

We have also worked hard at ensuring that direct assistance be provided to the laid-off workers. In conjunction with both provincial and federal politicians, we have laboured hard to ensure that the Rio Algom and Denison Mines adjustment committees, and more recently the community adjustment committee, were provided the necessary funds to carry out their work. Our greatest success in this area has been in the field of retraining. There are presently over 700 workers receiving, or about to receive, academic upgrading or skills training.

Our third area of concern, as community leaders, has been to develop vision and provide leadership in pursuit of economic diversification.

Even before the present layoffs were announced, the town had been working on an economic diversification strategy with the assistance of the Ministry of Northern Development and Mines. This document was completed in March 1990 and forms the basis of our activities. Initiatives have focused on three main areas: expansion of the retirement living program and related social and health care services; the retention and attraction of small business and government offices; and the attraction and development of training facilities within Elliot Lake.

Beyond the development of this strategy, however, our council and the community as a whole have been active in its implementation. Immediately after the Rio Algom announcement, council formed a mayor's action committee, in conjunction with the mining companies, the unions and business. From this group was developed a series of twelve briefs for consideration by the provincial government. They were not unfocused pleas for help, but rather specific proposals for action.

We wanted to demonstrate that Elliot Lake intended to save itself and wished government help, not handouts. In April we met with several cabinet ministers and presented our briefs. We have been successful in having the government form an interministerial committee for Elliot Lake headed by the Ministry of Northern Development and Mines and with representation from fourteen other departments.

Our economic diversification projects have met with some success. Our greatest achievement, of course, has been the retirement living program. Originally designed as a means to use up excess housing stock held by the mines, the program is now recognized by government as a major initiative in managing a "greying population." To date, over 1, 400 seniors have been attracted to the community, and a recently completed consultant's report has detailed a strategy to increase this number to 3,000 with an economic impact of tens of millions of dollars.

Other successes have included the establishment by the Ministry of Transportation of a regional training centre, creating fifty to seventy-five jobs, a commitment by the Ministry of Health to provide $1 million annually for ten years to operate a northern addiction rehabilitation treatment centre, and the establishment of a $10-million diversification fund for use by the community on projects of its own choosing. Numerous tourism initiatives have been undertaken which mean little in themselves but collectively represent an attempt to build a strong tourism and hospitality industry.

We have recently hired an economic development officer who will be travelling the province to encourage the relocation of secondary industry to our town. These are but a few of our successes and I am confident more will follow.

The survival of a community is not the result of one man, one idea and one initiative. Economic diversification is not simple, it is not fast, and it is not cheap. Employee adjustment is not painless, it is not easy, and it is not always entirely rewarding.

Meeting the challenge that faces Elliot Lake requires hard work, innovation and perseverance. I and my colleagues on council are fortunate, honoured and pleased to lead a community which has demonstrated, time and again, that it possesses the will, the dedication and the spirit to survive.

Chapter 9

Kirkland Lake

Joe Mavrinac

Kirkland Lake, like many other communities in Northern Ontario, has relied on the natural-resource sector, particularly mining and forestry, for its economic base. Unfortunately, these sectors are often sensitive to the whims of large-city corporate executives, fluctuations in senior government policies, designation, categorization, resource depletion, declining prices, uncompetitiveness and other variables which have spelled doom for many northern communities.

Our town, like the price of gold, has had its ups and downs. In 1938, at our peak, we had seven major gold mines in the "Mile of Gold," producing thirty million ounces. These mines directly employed some 4,000 or more miners, and the area's population was in excess of 25,000 people.

In 1954, after operating for forty years, Tough-Oakes, later called the Tolburn, became the first mine to close. The Wright-Hargreaves, which had produced between five or six million ounces of gold, closed in 1965. In 1968, after fifty years of mining and after producing close to ten million ounces of gold, the Great Lakeshore, Sir Harry Oakes's Lakeshore, shut down. On top of all this, the Upper Canada, located about eight miles to the east of Kirkland Lake towards Large Lake, closed in 1972. So the area had to deal with down-sizing in the 1950s, the 1960s and the early 1970s.

But the most difficult blow that Kirkland Lake had to endure was the province's infamous "design for development program" announced in 1966 after we lost the largest gold-producing mines in Canada. This stated very emphatically that the growth areas of Northern Ontario would be the largest communities in Northern Ontario: Timmins, North Bay, Sudbury, Sault Ste. Marie and Thunder Bay. For Kirkland Lake-no growth! It was because of our screaming that they changed that designation from "no growth" to "slow growth."

We had to do something in 1970, and we did: we hired an economic development officer and set up a commission for economic development what Elliot Lake is doing now. Out of the resulting development strategy came many wonderful things that helped all of Northern Ontario. These included the Kirkland Lake economic development assistance program, the first of its kind in Canada. They included the first core-storage library in Ontario. They included geophysical, geochemical and mapping surveys that led to the opening of the Holt-McDermott Mine two years ago. In addition, the area went into tourism. We refurbished and restored the Sir Harry Oakes's Chateau. We, the province, and the Oakes estate spent in excess of $1 million to create a beautiful facility housing the Museum of Northern History.

As indicated above, the 1960s found our major gold-producing mines closing their doors, with the exception of the Macassa mine, later to become Lac Minerals. That is the only gold producer that we now have. The number of employed miners declined to below 400 and the population plummeted and then levelled to about 13,000 people, a figure that still holds today.

In 1962, the Jones and Laughlin Steel Corporation of Pittsburgh gave us a tremendous boost when it announced it was opening the Adams iron-ore mine about fifteen kilometres southeast of Kirkland Lake, in Boston Township. The mine produced its first pellet in 1964, and in 1971 it was purchased by Dofasco of Hamilton, becoming a wholly Canadian-owned mine and employing some 400 workers, the vast majority of whom lived and did business in Kirkland Lake.

The announcement of the closure of the Adams mine in 1989 came as a surprise, especially in view of the fact that town officials had met the Adams mine management six months earlier and were told that the facility had seventeen years' more reserves which could be mined economically. Dofasco's slogan is "our strength is in our people." I found out about the closure when David Ramsay, our MPP, phoned me half an hour before the public announcement.

About the same time, the bubble burst on a very active mining exploration boom. Through a variety of causes, from higher taxes to special capital gains rules and the elimination of flow-through share tax breaks, junior mining companies were not able to raise exploration funding. In 1987, for example, the leading diamond drill company in Kirkland Lake had twenty-four rigs doing exploration work. In 1990, it had three in operation.

It is obvious that when a resource industry shuts down or becomes depressed, a municipality such as Kirkland Lake must bear the brunt of very significant social and economic adjustments. The town's welfare budget has increased dramatically. In 1990 we experienced a $300,000 deficit. The price of real estate has fallen, housing starts have dropped dramatically, and some business closures have taken place, particularly in the downtown core. The unemployment rate is approaching the mid-20-percent level.

OVERTURES TO GOVERNMENT

My first action in responding to this downturn in the mining sector was to seek help from senior levels of government. I commissioned a brief entitled "A Strategy for Survival for the Town of Kirkland Lake." This strategy was presented to David Peterson, then premier of Ontario, by the town of Kirkland Lake on 17 April 1989. The brief recommended that the province revise Kirkland Lake's designation as a "no- or slow-growth" area. It argued that the designation of only five growth centres in Northern Ontario created an environment suitable for out-migration of provincial jobs to the designated growth areas. This policy had a deleterious effect on potential entrepreneurs locating in Kirkland Lake, on tourism and even on the attitude of the local citizens.

The brief suggested that a strong provincial presence be provided immediately in the community in the form of a provincial building, not only to house those existing government offices which are totally inadequate, but also to relocate new jobs to Kirkland Lake. Such an initiative would provide confidence to outside investors and serve as a stabilizing factor. In the recent past, the Ministry of Northern Development and Mines relocated to Sudbury, Correctional Services to North Bay, and everything went to Sault Ste. Marie. What happened to Kirkland Lake? Here we are. We have a population of 13,000 people, with a community infrastructure that still can handle 25,000 or more people, because we have more services.

The brief also argued for an accelerated and innovative way of approving projects that would assist Kirkland Lake with economic development. For the long term, the brief requested that the province make a local government study of Kirkland Lake and its surrounding townships. The town believes that with a broader geographical base, it would be better able to deal with the needs of the area and would be more capable of implementing a development strategy. The brief requested that the Ministry of Municipal Affairs, under Bill 159, act within its authority to designate municipal assessment for the town of Kirkland Lake in a territory without municipal organization similar to that provided to the townships of Marathon and Manitouwadge so that the Holt-McDermott Mine would pay taxes to Kirkland Lake.

In addition to the brief to the premier, many other overtures were made to senior levels of government, both directly by the town of Kirkland Lake and through organizations such as the Temiskaming Municipal Association, the Association of Municipalities of Ontario, the Federation of Northern Ontario Municipalities and others.

On 26 July 1989, at the Temagami-Temiskaming Coordinating Committee, I made a presentation outlining the need for action regarding the economic and social well-being of the town of Kirkland Lake. Some of the requests I made were similar to those contained in the brief to the premier. But in addition I emphasized the importance of reforming the system which almost precludes municipalities from taxing mining companies in their jurisdictions.

At the time, I owned the largest motel in Kirkland Lake. I got very, very interested in what I had to pay when market-value assessment came in place. I found out that I just paid a few thousand dollars less than the largest gold producer in Kirkland Lake. I also found that in 1989, the largest gold producer, as reported in its financial statement, paid $9.5 million to senior levels of government, and $204,000 to the town of Kirkland Lake. How do you run a community with that kind of taxation structure in place?

The Northeastern Ontario Municipalities Action Group, of which Kirkland Lake is a member, prepared a document entitled "Strategic Issues for the Highway 11 Corridor," and we made that submission in 1989 to the six northern cabinet ministers of the new government. Some of the suggestions made in the brief were: that the province recognize the Highway 11 corridor as a viable planning unit; that it acknowledge the prime role that the transportation infrastructure plays in the development of the corridor; and that it assist the Coordinating Committee in identifying and analyzing the economic and social impacts of further highway improvement.

Some of our lobbying has paid off. Plans have now been completed and tenders are to be called for a secure custody facility for young offenders. This project will cost in excess of $4 million and will create about thirty permanent jobs. The Ontario Northland Transportation Commission now has under construction a multimillion-dollar bus terminal with 28,000 square feet of office space in downtown Kirkland Lake. The cost of this project will be $4.4 million.

The federal government has announced that an area office for the goods and services tax (GST) will be located in Kirkland Lake. The Ontario regional office of the Federal Department of Veterans Affairs is being relocated from Toronto to Kirkland Lake, and will provide 180 jobs by 1993. It already has obtained temporary quarters in a local office building, providing fifty jobs. Letters of interest and proposals are now being received by the federal government for the construction of a new 40,000-square-foot building to permanently house the Department of Veterans Affairs regional office and the GST office.

PRIVATE OVERTURES

We have not, however, relied totally on the senior levels of government for our recovery. Through our economic development department we have been actively pursuing new industries, businesses and institutions for our community. To this end, we have been somewhat successful.

We started negotiating with Northland Power about four years ago for a generating plant, now in its final stages of construction. It should go into full production shortly.[1] This is a $90-million project that will generate 105 megawatts of power, creating thirty-five to forty permanent jobs. Since Northland Power started its construction, ICG Utilities had to spend $4 million looping its gas line to the south end of Main Street, Kirkland Lake, because Northland's plant will be using seven billion cubic feet of natural gas per year. That's a major event.

Tim Horton's Donuts opened in January 1990, and McDonald's Restaurant in the summer of that year. That means nothing to southerners who have so many fast food places cluttering up their communities that they don't know what to do with them all. But the opening of these franchises in Kirkland Lake was very significant because these deals were made before the

Adams Mine announced that they were going to close their operation, and it took many trips by myself and other people to Toronto to convince people that we did have enough future left in Kirkland Lake for them to go ahead with their projects.

SOLID WASTE

The most important development, and probably the most interesting and most controversial one, is the proposal for Metro Toronto to use the Adams Mine site as a solid waste disposal facility. We are in the final stages of negotiations, and hopefully the agreement will be signed soon.

When this proposal was first presented to us we were adamant that it had to be a long-term project, include an industrial park for future development, have a recycling component, and have a research and development component. This whole package, over twenty-five years, could mean half a billion dollars to the community.

The biggest drawback to development of industry in Northern Ontario is transportation. We're working out an agreement with the rail companies that I'm sure is going to be resolved in the next couple of weeks.

CONCLUSION

Realizing that Kirkland Lake is going through a period of transition, we are greatly concerned by the sharp decline in mining exploration activity in our area. We in Kirkland Lake have long prided ourselves on our "Mile of Gold." Mining is still the backbone of our economy. Mineral exploration and development must be encouraged.

Because of this, mining-related communities of Northern Ontario and northwestern Quebec are banding together with the goal of encouraging the federal government to introduce a stimulant that would encourage exploration investment in order to bolster their seriously lagging local economies.

As in every other community that is single-resource based, diversification is our long-range plan. We cannot resolve the problems of a single-industry community such as Kirkland Lake with a one-pronged attack. Rather, I believe it will take a multitude of efforts in every possible direction, but conducted in such a manner that they are coordinated with all three levels of government, the private sector and everyone else who is willing to help us fulfil both our short- and long-term plans.

Notes

[1] Editors' note: On 25 May 1991, the *Globe and Mail* announced that Northland Power would be producing electricity in June of 1991. The plant is innovative in that it is fuelled by waste wood products and is considered environmentally friendly.

Chapter 10

Temagami

Ron Prefasi

Temagami is a community to which the closing of a mine comes as no surprise. The following is an excerpt from a 1985 editorial in the *North Bay Nugget* entitled "The Leading Edge":

> Is it proper for any municipality to pull a substantial chunk out of its reserve funds and go ahead with a massive, perhaps risky development scheme?
>
> That's exactly what the Township of Temagami is doing. In fact, within a short period of time, they will be signing the agreement which trades those reserve funds for a prime chunk of real estate along the waterfront of Lake Temagami.
>
> That waterfront may take three years or more to develop.
>
> Without meaning to panic anyone, Reeve Prefasi explains that everyone in the area realizes that Sherman Mine, which is the current economic anchor for a large chunk of this entire area, will eventually close. That's going to leave an enormous gap to be filled by other revenues. Failure to fill that gap is simply unthinkable – the consequences should be obvious. Therefore

> we must begin banking on the future of tourism to pull us out of a potential mess – before that mess becomes too big to deal with.
>
> Yes, indeed, Temagami council has the right idea – to deal with a problem before it gets out of hand.
>
> We praise them for having the courage to invest "money in the bank" for a real chance at a better future.

I know that my municipal colleagues would be able to share with you many such editorials about their own communities.

The point is twofold. Firstly, municipal politicians in the North don't "react" to the closing of a mine: we spend much of our political life fighting for the very existence of our communities, and when we have the luxury of a mine, we use that luxury to try to upgrade infrastructure to the point where we not only survive, but become attractive to investors or to tourists. It is the gravy from a mine that allows us to remain sound enough to live the northern lifestyle which we seek.

Secondly, the closing of a mine doesn't come as a surprise. There is an old saying which goes something like this: "The day a mine opens it begins to close." Nonetheless, a closing, though expected and not surprising, is still a shock to the system. An analogy might be that the day we are born, we begin to die, and know very well that it will happen; yet, it does come as something of a shock when we start pushing up daisies.

There is no doubt that the history of mining in Ontario is one of "boom and bust," a history which forces us to down-size every few years, and to expand when times are good. This is not the case in Temagami. We have not had the luxury of new mines opening on the heels of old ones closing, and even if that were the case, it could no longer happen due to native land claims. The mining industry in Temagami has not really been down-sized, as described in the given topic: it has been totally eradicated.

Temagami was not an organized municipality when Dofasco arrived to set up the Sherman Mine in the late 1960s, and therefore it did not really benefit in the ways that at one time were associated with a mine coming into a community. Our community did not get the recreation centres and the swimming pools or curling clubs. The one thing which it did get, however, was an increased tax base which gave it a boost into the future.

Temagami's municipal politicians have been preparing for that future and for the loss of the Sherman Mine for a long time. With Temagami only having become a municipality fourteen years ago, it lacked many of the hard services which larger communities sometimes take for granted. Even in the early 1980s, it did not have water and sewage systems serving its business community or its residents in the downtown core. Most township roads were rutted stretches of gravel. Temagami's waterfront, a privately owned piece of real estate, was a collection of tumbledown boat storage sheds, with docks which were no longer safe to walk on. The downtown core had begun to deteriorate to a point where the future of the business community was not looking bright.

In spending tax dollars from the Sherman Mine to really get the community into gear, Temagami's municipal politicians faced opposition from some local residents. Miners, often transient, were not really keen about politicians spending their tax dollars on infrastructure from which they would not directly benefit. They expected to be gone in a few years. They felt tax dollars from the mine should be used to keep residential taxes down.

Certain business people in the community felt they were making a good living before the mine came, and would continue to do so after it was gone, so why spend money on preparing for a future which would not be so bad anyway? Other people were afraid that they would not be able to support the community infrastructure being put into place once the mine went and insisted that council not do anything drastic, in fact, that it not do anything at all.

As well, seniors in the community wanted to retire there but feared that high taxes in support of new endeavours would choke their very existence. These were some of the greatest problems encountered by a council trying to spend today's dollars in order to stave off the problems associated with the eventual closure of a mine.

In spite of these and other difficulties, a number of actions were taken in Temagami over several years to prepare for the mine closure. The township bought the waterfront, upgraded the docks, rebuilt a 5,000-square-foot building to create a market square, constructed a new 12,000-square-foot building with a library, a ninety-five-seat theatre, a permanent display put together by the Ministry of Natural Resources, and a community meeting room. The township attempted to annex the islands on Lake Temagami which benefit from the municipality while not paying to help in its upkeep. It gathered $500,000 together to build phase one of an airport. It attracted $450,000 in funds through the Programme for Renewal Improvement Development and Economic Revitalization (PRIDE) to begin upgrading the downtown core, and it built a new senior citizens' housing project to keep our seniors in town.

In fact, the township did just about everything in the "future think" book.

We thought we were safe. We thought we were on the right track to ensuring a sound future. We thought we would have the time to get it all together before the mine closed. As it happened, we did not; and the premature closing of Sherman Mine brought its own problems to our small community.

It is difficult to dive into the quagmire of the last few years in Temagami and really separate the mine closure from other events around it as a cause of our present situation. Temagami has been a microcosm of the outside world, with influences on its future which have varied from a native land claim to an Ontario crown corporation purchasing its lumber mill in order to close it down.

Some days I can assure you that our future is definitely not in our own hands. Of one thing we can all be very sure: when disaster strikes in Ontario, we are not alone, and in fact, there have been some positive initiatives taken by the governments of Ontario and of Canada to help Temagami in its plight. Whether they are successful will be for history to judge.

For instance, in October of 1988, the province announced three initiatives which would help to guarantee Temagami's survival. Through the Municipal Economic Development Agency (MEDA), we were given $1 million to staff an economic development agency to provide full-time, long-term planning for the Temagami area. The government would give priority to present and future projects for the Temagami area. This meant their acceleration or enhancement to provide additional economic impetus to the area. Special capital initiatives for the area, which could range from tourism-based to resource-oriented projects, would also be developed and brought forward.

Further, again in 1988, the province set up a group called the Temagami Coordinating Committee, a collection of bureaucrats from various ministries. It was organized to streamline the interrelationships of ministries, thus helping projects in Temagami to move more quickly, and probably would have been of immeasurable help. However, when the mine closures were announced, the group grew to more than thirty bureaucrats in two divisions (social and economic) and became the Temagami/Temiskaming Coordinating Committee. The committee has collected a vast amount of socio-economic data on the area to date, and has spent innumerable hours, days and months discussing the area's situation. However, I personally think the committee has been given a very difficult, if not almost impossible, task.

In 1989 the province sent in a team led by the Ministry of Tourism to develop a tourism strategy for Temagami. In that same year, a development group called Temcor set out a proposal for vacation condominiums in a destination resort with ski hills and golf courses, and was granted money for a feasibility study.

In February 1990, the Ministry of Northern Development and Mines (MNDM) announced the coming of Pyrok, a plant which would produce a new type of wallboard with cement and wood particles. Pyrok would set up on part of the Sherman property and employ up to seventy persons within the next five years.

Let me now turn to the federal government incentives. The Section 28 plan in our area has been increased by an incredible number of dollars. This is an unemployment insurance plan which allows us to hire unemployed persons on UIC and enhance their benefits. A good idea; however, it has its drawbacks. One is that the program does not give dollars for materials, and the second is that we are unable to find people who wish to take part in the program.

What of our corporate citizen Dofasco? The Sherman Mine had a very high profile in our community and in our area for over twenty-five years. Today you won't find anything in Temagami to remind you of its presence. It has disappeared. Temagami made a number of approaches to Dofasco before the Sherman Mine closed, offering a partnership in our future in terms of shared funding for a Sherman Park; asking that the land which has been held by the mine as a buffer zone around its operations be turned over to the community; and requesting help in developing our airstrip, which would be located on part of the tailings basin at the mine. In discussions with Dofasco's young media executives, we were informed that in the future Dofasco could not be counted on for maintenance funds for parks or whatever was then being done in partnership with the town. This was quite an attitudinal shift from a few months earlier, when the mine was involved in everything in Temagami, from ski hills to cemeteries!

Now, I don't have a scorecard, but at the time that all of these private and government announcements, initiatives and plans were made, and when all of these ideas were put into motion, we had an operating lumber mill with 136 employees, and a mine with over 300 employees. Today, both are gone, and nothing has replaced either one.

That is a short, and very incomplete, look at our past. Here, very quickly, is Temagami today.

The province has stopped our annexation proceedings because of land-claim negotiations with the Teme-Augama Anishnabai; the question of the claim has been with us, unresolved, since 1973. Dofasco will not allow the town to use a small portion of the tailings pond for an airstrip unless the town assumes all environmental responsibility for whatever might surface in the tailings pond in the future. Dofasco is also preparing to sell individual lots piecemeal to residents. The Ontario Development Corporation has purchased the Milne Lumber Company, has closed the operation, is selling off the assets and is in the process of looking to dispossess twenty-three families living on the property. Temcor is starting over, trying to convince a new government that it has a worthwhile proposal. Pyrok has disappeared into the proverbial woodwork.

In spite of all this, we will survive as a community. Not only will we survive, we will become attractive to tourists and eventually to investors so that we are able to grow into a well-rounded northern community with a stable tax base and a solid economy. When the dust finally settles, we'll be okay: that is, if we haven't destroyed ourselves through the infighting which occurs whenever changes are made, and if the New Democratic Party doesn't settle native land claims by giving away all of our prime real estate, and if the governments of Ontario and Canada continue to support our efforts for diversification. Certainly there are a lot of "ifs." We hope to turn each one into a "when."

Without land for development, and with an unsure future staving off investment, it will not be easy to turn our economy around. With investors, ideas, perseverance and a lot of luck, we can be successful.

Part Five

Into the Future

Sudbury Regional Development Corporation

Opened in late 1970s, and set alongside newly built municipal government offices, the "Provincial Building" symbolized government commitment to a revitalized Sudbury.

Introduction

Part Five of this volume makes it painfully clear that the fortunes of Northern Ontario's mining industry continue to be unstable, and that the future of the region's communities is also uncertain. For many towns and cities, the answer is economic diversification. Oiva Saarinen's "Creating a Sustainable Community: The Sudbury Case Study" is a timely analysis that shows how a relatively small and dependent mining community, faced with the challenge of a shrinking mining workforce, was able to transform itself into a regional economic powerhouse. Saarinen, a Laurentian University geographer, offers this experience as a model that might well be studied by other communities in the North and elsewhere, not only for inspiration but for specific ideas. Wallace would be pleased.

And yet Saarinen argues that there is a critical mass of population necessary for the kind of phenomenon that took place in Sudbury to replicate itself elsewhere. Unfortunately, few other Northern Ontario communities possess anywhere near this critical mass. This also means they lack the political clout needed to attract federal and provincial largesse – largesse that in Sudbury's case took the form of a taxation data centre, the offices of the Ministry of Northern Development and Mines, and the like.

In this context, Mark Shrimpton and Keith Storey's paper on "Fly-In Mining and the Future of the Canadian North" takes on added significance. For Shrimpton and Storey, both of whom are geographers associated with Memorial University's Institute of Social and Economic Research, the era of resource-based, single-industry towns is over. Instead, in the future, mining companies will fly their workers in to mine sites for just as long as it takes to exploit the mines. Because there will be no communities set up to support the mines, there will be none left over to support when the mines close. All of this raises some interesting questions. Are communities like Elliot Lake, Kirkland Lake and Temagami, now populated with large numbers of qualified but unemployed miners, doomed to disappear? Or could *fly-out* mining be one of the keys to these communities' continued existence? At least one of the three municipal leaders represented in this collection, George Farkouh of Elliot Lake, has already incorporated the idea into his community-survival strategy. Time alone will tell what fate awaits these communities.

Chapter 11

Creating a Sustainable Community: The Sudbury Case Study

Oiva Saarinen

Sudbury serves as a historical case study of a resource-based settlement that has undergone several transformations since its inception as a village in the latter part of the nineteenth century. This paper suggests that Sudbury is unique among Northern Ontario communities inasmuch as its evolution has involved four major stages of change:

1. from a railway village to a colonial-frontier mining town and city
2. into a regional central-place
3. to a declining metropolis and finally
4. towards self-sustainability

A number of factors can be identified which have contributed to the restructuring of its economic base and the reshaping of its society and culture. In this process, the "human dynamic" has been a vital contributor to change. The paper is positive in tone as it suggests that under certain circumstances a resource community can progress from a "boom-or-bust" dependence upon staples to a more sustainable form of urban economy based on new local and regional roles.

FROM RAILWAY VILLAGE TO A COLONIAL-FRONTIER TOWN AND CITY

Map 11.1 Geographical Aspects of the Sudbury Area in the Colonial-Frontier Era

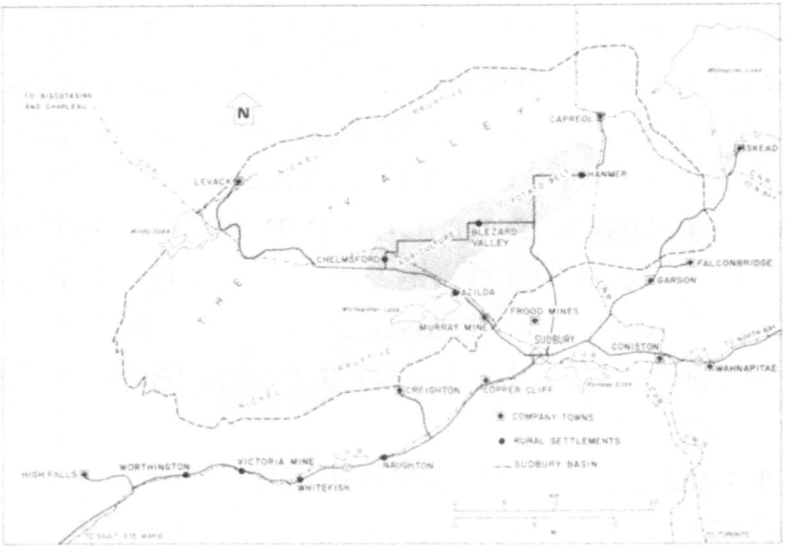

Source: O.W. Saarinen.

Sudbury's first transformation was from a small Canadian Pacific Railway (CPR) townsite in the 1880s to a full-fledged mining community by the turn of the century.[1] In this early "staples" era, the export of nickel and other minerals set the pace of economic growth and left a strong imprint on community development.[2] By World War I, the Sudbury area reflected many features of what can be called a "colonial-frontier" setting, including resource dependency, environmental pollution, external control and an uncertain future, a low quality of life and a corporate paternalism in the form of the company town. These characteristics combined to give the Sudbury area an extremely negative image throughout Canada.

Sudbury began life in 1883 as a company townsite of the CPR. Originally conceived as a temporary townsite, perceptions of its future changed drastically between 1886 and 1888 when mining and smelting operations were started in nearby Copper Cliff.[3] By then it was clear that the village's location, beside the geological structure known as the Sudbury Basin, had effectively sealed its fate as a mining-based community (Map 11.1). The introduction of mining necessitated the creation of roast yards and slag and tailing zones, which over the years fostered bleak images of the area as being similar to "death valley" and to the "surface of the moon."[4] The staples economy came to be influenced by a series of external impulses which perpetuated the region's "hinterland" status.[5] By 1902 Inco had been formed with its headquarters in New Jersey, and thereafter, the perception of Sudbury as an "Inco" town was firmly rooted in the public mind.[6] While Inco nominally became a Canadian operation in 1928, real corporate power continued to reside south of the border.

The resource dependency of the Sudbury area made it highly susceptible to the influence of the international nickel market (Figure 11.1).

While fluctuating market conditions had little impact on the overall growth of the population, they nevertheless contributed to a feeling of uncertainty about the community's long-term economic future. Sudbury also long remained isolated from southern Ontario. Only in 1908 and 1909 were direct railway links forged to Toronto.[7] This enabled the latter to extend its sphere of influence at the expense of Montreal, which had had a head start because of its initial connection to the region. Within the developing Northern Ontario highway system, Sudbury was at first peripheral to North Bay, which became the major regional crossroad following the construction of Highway 17 west to Sudbury and Sault Ste. Marie in 1912. This pivotal gateway setting gave North Bay a powerful advantage over Sudbury until the 1950s.[8] The external dependency of Sudbury was reflected in the provincial government, whose policies as they affected the Sudbury area, served only to attract foreign capital and to divert development benefits from local treasuries. It was clear that the negative side-effects of the development process were to be borne by the local residents and communities alone.[9]

The staples economy and external dependency of the Sudbury area had a major impact on urban development and community life. A number of dispersed centres evolved in the area, many of them company towns (Map 11.1). These included the Inco communities of Copper Cliff, Creighton, Coniston, Levack, Garson, Frood Mines, Murray Mine and High Falls. The company town was also introduced by Falconbridge Nickel following its creation in 1928.[10] Company towns were constructed as late as the 1950s at places such as Lively and Onaping.

Another feature affecting urban development was the fact that the mining companies did not pay property taxes like other industrial operations in Ontario. The early provincial policy that mining companies were struggling firms that should not be required to pay municipal taxes was later incorporated into the Assessment Act. This situation created hardships for Sudbury, which lacked any alternate form of industrial assessment. This policy was partially rectified by the "mining-revenue" payments introduced by the province after World War II; nonetheless, these monies were equivalent to only one-half of the industrial revenues received by a typical southern Ontario municipality.[11] One study of this problem has concluded that the special tax treatment accorded the mining industry was a major factor leading to underdevelopment of physical infrastructure and a low quality of life in the Sudbury area.[12] Continuing negative references to Sudbury in the national press gave testimony to the legacy of this provincial assessment policy.

Figure 11.1 Population, Employment and Nickel-Production Trends in the Sudbury Area (1901–1990)

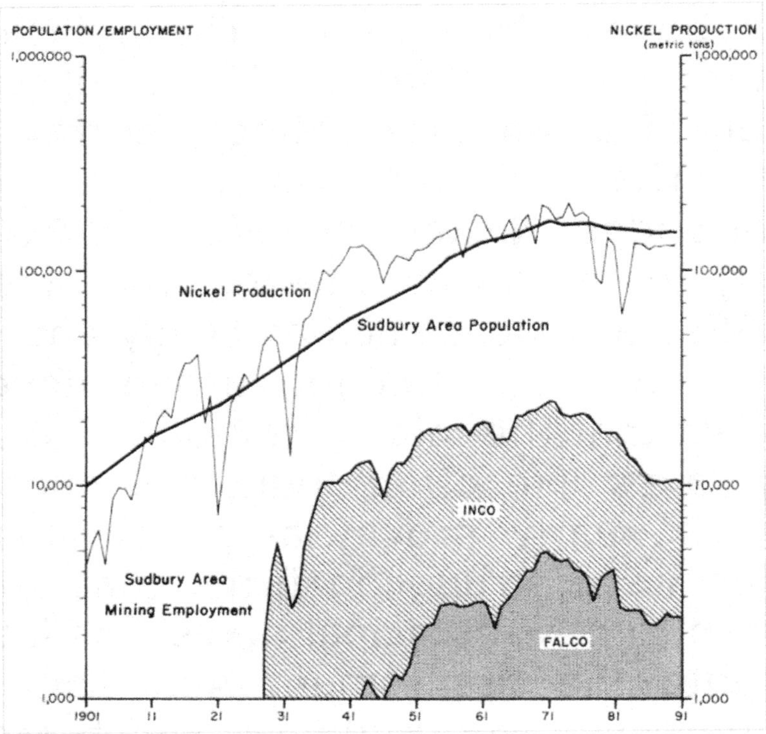

Sources: Ontario Ministry of Northern Development and Mines, Ontario Mineral Score, Toronto; Statistics Canada, Census of Canada, 1901-1906, Ottawa; Regional Municipality of Sudbury, Regional Planning and Development Department; and data provided by Inco Ltd. and Falconbridge Ltd.

The colonial-frontier setting exhibited several distinctive socio-cultural features. The company towns, for example, reflected the reality of segregation by class and race.[13] In the Sudbury Basin company towns, fringe settlements occupied by immigrants developed. Sudbury itself served as a regional "fringetown" for both the Inco and Falconbridge company towns. Its population figures confirm this role. In 1901 Sudbury embraced 21 percent of the area's population: even as late as 1956, only 40 percent of the region's inhabitants could be found within its boundaries. The company-town phenomenon had a great impact on Sudbury's internal development. The company towns attracted virtually all of the professional employees of the mining industry, leaving Sudbury without a representative middle and upper class aside from a token number of doctors, lawyers and clergy. This skewed class structure shaped the character of local politics and fostered a blue-collar dominance of community tastes and demands.

TOWARDS A REGIONAL CENTRAL PLACE

After World War II, Sudbury began to shed aspects of its colonial-frontier character. Nickel demand grew in response to military and consumer needs. As the long-term future of the Sudbury Basin seemed assured, concerns regarding community instability waned. This postwar stage, however, did not lead to a shift from local staple production to a diversified mining base featuring locally manufactured inputs and diversified products.[14] Rather, the Sudbury area provided support for the contention that staple economies generally develop more of the same rather than evolve into diversified industrial complexes.[15] Nevertheless, the unique size of the Sudbury Basin mineral deposits, the great boom of the 1950s and the attainment of a metropolitan status encouraged import substitution and the expansion of business and consumer services throughout the region. This metamorphosis was underlaid by a fundamental strengthening of Sudbury's regional status and the growth of its hinterland in northeastern Ontario. The growing maturity of the Sudbury community was reflected in its acceptance of land-use planning, the diversification of its physical infrastructure and the strengthening of its middle and upper classes.

The evolution of Sudbury from a colonial-frontier community to a central place in northeastern Ontario was initially the result of direct intervention by the American military-industrial complex. Concerned by the Korean and Cold War, and the global expansion of communism, the United States government encouraged greater international competition and production for strategic minerals such as nickel.[16] Due to its strategic mineral potential, the Sudbury Basin received $789 million between 1950 and 1957 to diversify the nickel supply and other metals through stockpiling and special purchase agreements.[17] In the Sudbury Basin, Falconbridge Nickel proved to be the major beneficiary of the American initiative. By the end of the 1950s, Falconbridge had risen to the rank of a dominant world producer through an aggressive mining program focused on the northern part of the Sudbury Basin.

The local consequences of this mining boom were profound. The labour force grew rapidly and so did the population. In 1951 the regional population exceeded 115,000; by 1961 it was 138,000 and by 1971 around 170,000.[18] Sudbury thus managed to acquire the unique distinction of being a mining-based metropolitan area. Crossing this metropolitan threshold made it possible for the local economy to expand via import substitution and the growth of health, education, business and consumer services. Important in this connection was the addition of three hospitals between 1950 and 1956 and the creation of Laurentian University and Cambrian College in 1960 and 1966 respectively.

In the meantime, two other influences contributed to the growth of the local service sector. The first was the construction of Elliot Lake west of Sudbury from 1956 to 1959. This new community of 25,000 provided an expanded market for retailing and wholesaling. The second factor was generated by the opening of the Sudbury-Parry Sound-Gravenhurst highway between 1952 and 1956, which gave Sudbury a direct road link with southern Ontario.[19] For the first time in its history, Sudbury acquired a "gateway" location, and this allowed it to evolve as a major competitor with North Bay.[20] These two developments enhanced Sudbury's influence in the region and paved the way for a broadening of its hinterland.

This extension of Sudbury's sphere of influence is evident in newspaper circulation figures which reveal that the *Sudbury Daily Star* supplanted Toronto newspapers in Manitoulin Island in the 1950s.[21]

The opening of Canada's first privately owned television station served as another indicator of Sudbury's growing importance as a regional centre. A national dimension was added with the introduction of Trans-Canada Airlines flights in 1952-54. Through these flights a firm link was established with the air corridor that has been called Canada's "main drag."[22]

The economic maturation of Sudbury was accompanied by changes throughout the region and within the city. Previously, the urban structure of the region consisted of self-contained communities, with Sudbury serving essentially as a local centre for the Sudbury Basin communities. The changed distribution of mining employment in the 1950s, however, completely reshaped the urban setting. Urban sprawl occurred in the centre of the basin known as the "Valley" and along the fringes of the city. Studies have shown that these changes resulted in the formation of a functionally integrated urban system throughout the basin. Unfortunately, the system remained highly fragmented politically.[23]

A number of changes also occurred within the city. The laissez-faire attitude of the city government was grudgingly set aside and replaced by one favouring more controlled growth. This was evident in the hiring of a city planner in 1955, the establishment of an industrial commission and a conservation authority in 1957 and 1958, and the adoption of the city's first official plan in 1959.[24] Planning proved to be a boon, as shown by the successful urban renewal of the downtown, improved transportation linkages, the elimination of flooding, and the provision of water and sewer facilities to suburban zones. The community skyline reflected a new dynamism as old landmarks were demolished and new structures erected. Throughout the 1960s, additional employment opportunities in commerce, health and education gave the community a more powerful middle- and upper-class white-collar mentality.

By the 1970s, then, Sudbury was clearly in the process of shedding its former role as an Inco mining community dominated by blue-collar workers. While it is true that the Sudbury economy was still caught on the staples treadmill,[25] there was evidence that its role as a regional centre was becoming increasingly important.[26] This broadening of its economic base improved the quality of community life. By 1971 Sudbury was closer than ever before to offering a social and cultural environment equivalent to other Ontario cities of its size.[27]

These progressive changes, however, were not fully appreciated at the time due to the intense national and international media attention accorded the rivalry between Inco and the Mine Mill Local 598 in the 1950s and 1960s. Coverage of this rivalry fostered images of the Sudbury area as being a "hotbed of unionism and communism."[28]

A DECLINING METROPOLIS

The post-war transition came to an end during the 1970s and early 1980s when the Sudbury area faced what has been termed a "turbulent environment," involving challenges of a political and economic nature.29 During this period, new problems threatened the future viability of the city and the entire Sudbury Basin. The region not only faced major political and planning issues stemming from its fragmented and inefficient municipal structure, but was also plagued by an unprecedented downturn in the mining economy. The latter crisis attracted widespread attention in the press, and the impression that Sudbury had "hit bottom" and was "struggling to stay alive" rapidly dominated the public's image of the area.[30]

The urban crisis of the 1970s can be directly linked to the settlement sprawl which followed the economic boom of the 1950s and the inability on the part of the province to deal resolutely with mining-assessment. The first attempt to deal with this problem occurred in 1960 when the city of Sudbury was enlarged to incorporate the large fringe population found adjacent to its northern and eastern boundaries. This municipal reform, however, did little to deal with the lack of regional planning and the inequitable sharing of the mining assessment base among settlements. No major move was made in this direction until 1968 when the province's Design for Development policy called for the reform of local government structures.

In line with this reform, the Kennedy Report was issued in 1970, recommending a two-tiered form of regional government, and the equalization of the mining-assessment base.

The report led to two years of intense debate at the local level, leaving little doubt that no form of consensus was possible.[31] It thus remained for the province to intervene directly; under Bill 164, a new entity called the Regional Municipality of Sudbury finally came into being on 1 January 1973 (Map 11.2). This was a milestone in the political and planning evolution of the area because it provided the Sudbury Basin communities with an administrative framework for dealing with regional problems at the appropriate regional level. Sufficient controversy, however, remained between Sudbury and the outlying municipalities to delay the implementation of the Official Plan for the Sudbury Planning Area until 1978.[32]

The second aspect of the turbulent environment was the drastic decline in mining-based employment. The roots of this decline can be traced back to the 1960s and 1970s when Inco began to lose its dominant position as the world price leader for nickel. New suppliers appeared on the scene, forcing nickel prices downward. Faced with heavy investments in Guatemala and Indonesia, and hurt by the OPEC (Organizing of Petroleum Exporting Countries) crisis of 1973, Inco found itself burdened with high debts and nickel inventories. Cost-cutting measures were introduced between 1972 and 1974, resulting in the loss of 3, 700 jobs to the Sudbury area.

The economic situation worsened in 1978-79 due to a lengthy strike against Inco by the United Steelworkers of America Local 6500. The effects of this strike were devastating for both Inco and Sudbury. For Sudbury, it meant yet another resurrection of the "mining" and "union town" image. For Inco, it resulted in a further decline of its global market share and a negative public image. Already identified as a "corporate bum" in the federal election campaign of 1972, Inco had to face *Canadian Business Magazine's* charge that it exhibited "appalling" labour relations and colossal "arrogance."[33]

Map 11.2 Site and Setting of the Regional Municipality of Sudbury

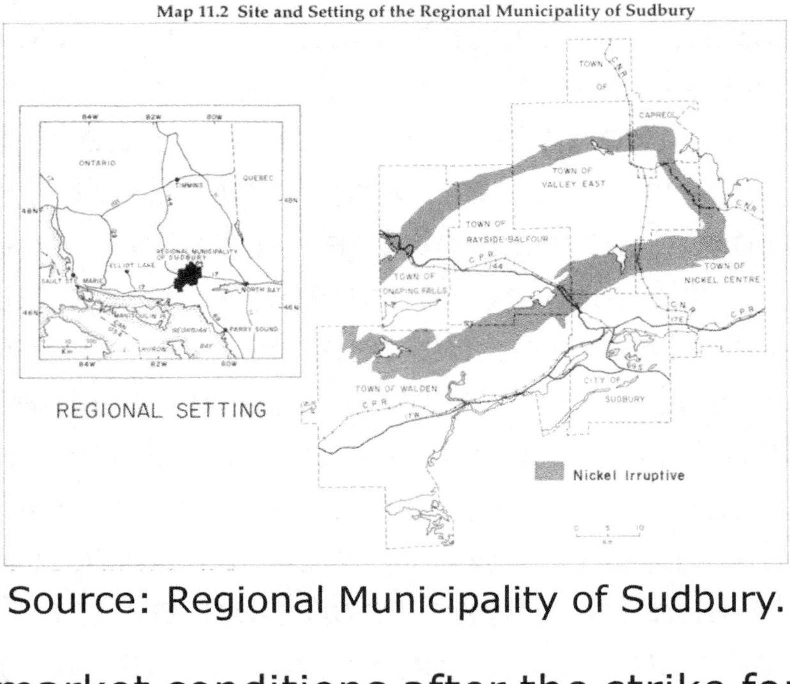

Source: Regional Municipality of Sudbury.

Poor market conditions after the strike forced Inco to reduce its labour force by another 7,000 during the 1980s. Falconbridge Nickel managed to retain most of its labour force until 1978, after which it was forced to follow Inco's lead. The Sudbury area, therefore, experienced a drop in its mining-based employment level from a historic high of 25, 700 in 1971 to 17, 700 by 1981. By this time a realization had dawned within the area that the glory days of the mining era had come to an irrevocable end. This new reality was confirmed by the fact that from 1981 to 1988 mining employment dropped by yet another 7,000 before stabilizing at the 10, 500 level.

This drastic downturn of the mining economy had a great impact on population levels in the area. In 1971 the Regional Municipality of Sudbury had a population of 170,000; by 1976 and 1981 this figure had dropped to 167,000 and 159,000. Consequently, Sudbury gained the dubious distinction of being, along with Windsor, one of only two metropolitan areas in Canada to experience population loss between 1971 and 1981. This pattern of decline continued until 1986, by which time the regional population had dropped to 152, 500.

MEETING THE CHALLENGE: TOWARDS A SUSTAINABLE COMMUNITY

The twin crises associated with the "turbulent" era did not go unchallenged. Following an agonizing period of reappraisal, a positive response was organized which eventually resulted in the rebirth of a much different Sudbury. This response proved to be so effective that as early as 1984 Sudbury was selected by government agencies, such as the Organization for Economic Co-operation and Development (OECD) and the Ontario Ministry of Municipal Affairs and Housing, as an international model of successful urban-economic adjustment.[34] By the middle and late 1980s it had also become clear to the Canadian investment community that, despite its population decline, the Sudbury area had one of the most stable economies in the country.[35] More recently, Sudbury's revitalization has become apparent to the general Canadian population because of favourable articles in *Chatelaine* and *Maclean's* magazines.[36]

How did this positive response come about? A number of factors can be identified that combined in synergetic fashion to halt economic and population decline and stimulate recovery. Many of the preconditions for this adjustment – the development of a metropolitan threshold, the expansion of central-place functions and the development of a solid middle-class and white-collar power base – were inherited from the post-war era. These three long-term factors were complemented by four newer impulses: creative political and planning leadership, financial assistance from the senior levels of government, managerial and productivity improvements put into place by Inco and Falconbridge, and finally, the broadening of the mining-related sector and the related promotion of Sudbury as a world-class technological centre for mining.

In assessing the dynamics of the adjustment process, that of community size is of paramount importance. In 1986 the Regional Municipality of Sudbury still had a population in excess of 152,000. While this figure did not approach the 250,000 mark, the population size often considered necessary for self-sustainability, it nevertheless proved sufficient to serve as a brake on the process of decline. Operating under the dictum that "no nation is so affluent that it can afford to throw away a major city," civic boosters and local politicians lobbied both the provincial and federal governments for assistance.[37] These attempts, discussed later, proved to be highly successful. Another feature linked to the population threshold was the opportunity it afforded for import substitution. Even the downturn in employment and population levels had some positive effects, such as the lowering of housing costs and the growth of the mining-retirees sector. Retired miners now outnumber those employed by Inco and Falconbridge.

Throughout the 1970s and 1980s, the Sudbury area strengthened its role as a regional centre in northeastern Ontario.[38] This status was reinforced considerably by improvements in transportation and communications. In 1970, Highway 144, linking Sudbury to Timmins, was opened; it was subsequently extended to include Smooth Rock Falls and Cochrane via Highway 655. This allowed Sudbury to exert an influence in areas formerly dominated by North Bay. The flow of traffic within and around the regional municipality was improved through the construction of bypasses and arterial roads. With the erection of a modern air terminal in 1974 and the establishment of east-west air links to Winnipeg and Montreal, Sudbury's connections with the rest of Canada were solidified. As indicated by the success of the Walden Industrial Park and the opening of two large wholesale operations by Loeb and National Grocers, these improvements enhanced Sudbury as a distribution point for all of northeastern Ontario. Similar advances occurred in communications as demonstrated by the shifting of Bell Canada's district office to Sudbury from North Bay in 1978 and the upgrading of this office to that of a regional headquarters. In the areas of finance, insurance and real estate, downtown Sudbury has become the focal point for new construction and employment opportunities, including, for example, three northern corporate headquarters in the banking sector.

It was improvements such as these that underlay the transformation of the city from its localized role relative to the Sudbury Basin towards regional dominance within northeastern Ontario.[39] This status resulted in central-place functions becoming a dominant driving force within the economy. Support for this metamorphosis is given by research suggesting that Sudbury by 1981 could be considered as a "service-administrative" centre within the Canadian urban context.[40]

The third factor in the revitalization process involved the continued growth of the middle and upper class and the concomitant change in the urban mentality from Stompin' Tom's "Saturday Night" to a more sophisticated culture including live theatre and a symphony orchestra. The refurbished Grand Theatre attracted well-known entertainers from all over the world. After the early 1970s, political life was also reshaped by the election of representatives of a more professional bent.[41]

The planning process increasingly included quality-of-life considerations such as leisure facilities and the preservation of green space. Perhaps the most noteworthy example of the new mind-set was the creation of Sudbury 2001. This body, formed in 1978 and disbanded in 1987, had the objective of making the Sudbury area a self-sustaining metropolis by the turn of the century. Supported by a broad spectrum of local groups and backed by provincial funding, it embarked on a venture-capital program designed to diversify the local economy. It had a substantial impact in the forging of a longer-term planning perspective. The body also succeeded in replacing the traditional pattern of confrontation and divisiveness within the regional community with one based on pride, cooperation and unity. This long-term form of strategic planning eventually bore fruit, as other events in the 1980s demonstrated. One authority has suggested that the Sudbury 2001 experiment deserves serious consideration by other communities facing adjustment difficulties.[42]

The fourth transitional force involved the high level of political and planning creativity achieved by the Regional Municipality of Sudbury. Despite its shaky start in the 1970s, regional officials and politicians implemented actions concerning environmental improvement, economic development, land-use planning, fiscal planning, and emergency job creation.

One of the first steps taken by regional government dealt with the issue of environmental pollution. Stung by the "lunar landscape" image evoked by the visit of the Apollo training missions in 1969-70, initiatives were taken to form a tree-planting committee in 1973. High unemployment levels in 1978 and 1982 later provided the workforce necessary to implement a major land-reclamation program unparalleled in Canada. This "greening of the environment," which has continued to the present, produced spectacular results, causing the *Hamilton Spectator* to comment as early as 1984 that "birch trees, lakes and the grass – sprouting out of once-black hills – form a beautiful backdrop to the city."[43] Since then, more than one million trees have been planted.

Economic development emerged as another priority. In 1974, the Sudbury Regional Development Corporation (SRDC) was established as an autonomous body of representatives from business, industry and government, accountable to the regional council, but with its own staff and budget. Sudbury thus became one of the first municipalities in Ontario to recognize the need for an autonomous catalyst to promote economic development. In addition to promoting new employment, the SRDC has been actively involved in the broad area of image transformation and public relations.

An enhanced form of urban managerialism was also introduced. This attempt to increase the efficiency of regional government was manifested by the adoption, in 1984, of a declining debenture policy to reduce debt charges. It was also evident in the land-use policies of the Secondary Plan for the Settlement of Sudbury, approved in 1987, which called for residential infilling, land-use recycling and a "Jane Jacobs" type of diversified downtown.[44]

The massive layoffs of 1982 and the media image of Sudbury as a doomed city met with creative political leadership. Drawing on the experience of Sudbury 2001 and the belief that the two senior levels of government would respond best to firm political leadership with clear authority, the chairman of the Regional Municipality of Sudbury instituted a plan to deal with the short- and long-term future of the community. This plan consisted initially of an intense networking process involving the solicitation of public opinion and interest-group input. The result was an eight-part mission statement linked to mining, government, business, industry, finance, health, agriculture and education/training.[45] Discussions with senior provincial and federal officials tied the task force rec-

Figure 11.2 Canadian Census Metropolitan-Area Rankings by Value of Building Permits (1978–91)

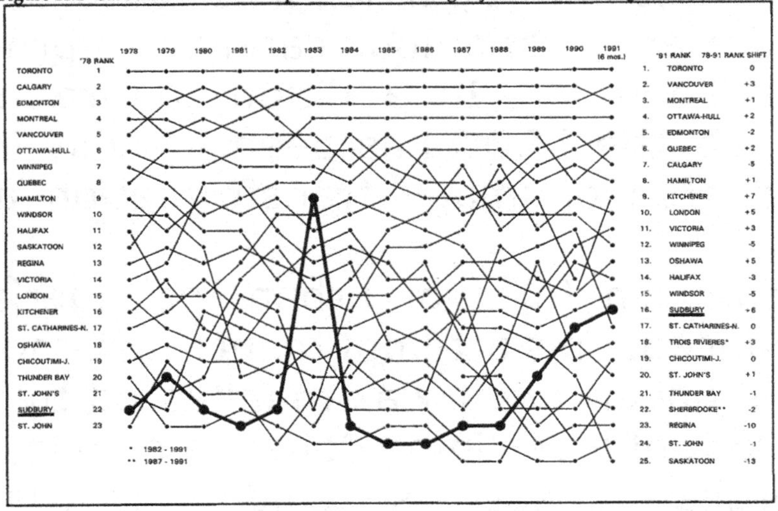

Source: Statistics Canada, Building Permits (December 1979-June 1991), Ottawa.

ommendations to existing and proposed programs and sources of funding. This strategy helped to cushion the short-term blow of the 1982 layoffs and laid the framework for the successful longer-term planning undertaken by the SRDC and the Corporate Plan of the City of Sudbury.[46]

The fifth element of the recovery phase was the direct investments made by the provincial and federal governments. The timing of these investments proved to be fortuitous, as they coincided with the contemporary trend to decentralize civil service bureaucracies from Toronto and Ottawa. This process started after the 2001 Conference in 1978 with the erection of a new provincial building in the downtown Civic Square complex. This lent a new dynamism to the city core and laid the framework for the emergence of Sudbury as a government centre. In 1984 Science North was opened, and its outstanding architecture helped give the area a new image. The facility fostered the development of a small but growing hospitality and tourism sector. In 1990 the head office building of the Ministry of Northern Development and Mines was opened, involving the relocation and/or generation of 250 jobs. In the same year the Northeastern Ontario Regional Cancer Centre was established, contributing another 250 jobs to the local economy and consolidating the region's position as a medical referral area. A similar number of jobs were created when the Mining and Minerals Research Centre at Laurentian University (housing the Ontario Geological Survey, the Miners' Health and Safety Centre, and the Ontario Research Directorate for the Ontario Mining Association) is opened in 1992. Not to be outdone, the federal government stimulated the local economy in 1982 when it established the Sudbury Taxation Centre, providing some 750 full-time and 1, 500 part-time jobs. Taken together these government investments have had a substantial impact on construction, the diversification of the economy and the fostering of white-collar jobs[47] (Figure 11.2).

The sixth transitional factor was the corporate response of the Inco and Falconbridge mining giants to the events of the 1970s. Increased global competition, the visual effect of the superstack on the environmental movement in North America, and the negative public relations impact of the 1978-79 strike together forced Inco to reassess its corporate image, efficiency and labour relations at the turn of the 1980s. The changed pattern of the market likewise forced Falconbridge to revamp its corporate objectives. Thus, both companies introduced productivity-improvement strategies incorporating the environmental ethic and better management-labour relationships during the 1980s.

Innovations such as mechanized mining equipment, automated systems, electrified operations, computerized controls and continuous mining operations were introduced. The effects were impressive; as the decade progressed the two firms recorded productivity improvements in the order of 70 to 100 percent. While these advances initially involved labour-force reductions, a temporary stabilization of the workforce occurred by the late 1980s; then Inco began to hire new miners in 1988 for the first time since 1979.[48] Both firms made major capital expenditures in environmental programs designed to substantially reduce sulphur dioxide emissions by the mid-1990s. The above events lend support to the theory that an era of economic turbulence is often followed by innovative corporate responses to a new institutional and technological setting.[49] While it is true that both Inco and Falconbridge have undergone many of the internal transformations required for the 1990s, recent events have raised the possibility of yet another round of productivity-improvement programs and some labour-force adjustments.

The seventh and final aspect of the revitalization process has been the belated emergence of a movement to make the Sudbury area a world-class mining-based complex. Recent advances in this direction indicate that the region is now entering the stage of a mining-based "export complex."[50] Both forward and backward linkages have appeared in the mining sector. Nickel products have been broadened to include forty to fifty nickel grades of powders and pellets; a rolling mill has also been established. Local manufacturing has been transformed by the replacement of former wholesaling and distribution branch services with fabrication and assembly plants producing a wide variety of machine parts and equipment. This has been fostered, in part, by the trend towards reduced inventories by the two companies, the adoption of continuous mining techniques, and the need for immediate servicing and parts replacement by local experts. The importance of Sudbury in this regard is reinforced by the fact that some ninety mines can be found within 300 kilometres of the Sudbury Basin.[51] This has induced a number of European mining-equipment companies, for example, Tamrock from Finland, to establish their North American headquarters in the area.

These linkages have been supplemented by a broader-based strategy for the transformation of the Sudbury area into an international mining complex featuring the use of the local mines and labour force as a technological testing base for the production and export of sophisticated mining equipment. In 1984 Inco reactivated its Copper Cliff North Mine as a full-scale mining research facility featuring automated equipment and computerized controls. Falconbridge similarly initiated a special Mining Technology Group using the Onaping Mine as a test facility. Continuous Mining Systems was created by Inco in 1984 as a company to design, manufacture and market innovative lines of mining equipment. A major step was taken by this firm in 1987 when it achieved its first export sales.[52] This represented a key shift in the move of the local economy from an "extractive" towards a more "technologically based" mining export complex. The establishment of the $55-million Mining and Minerals Research Centre, along with the existing Centre in Mining and Mineral Exploration Research on the Laurentian University campus, was another indicator of the evolution towards such a complex.[53]

Evidence that progress was being made towards the vision of a self-sustaining community came in the late 1980s with the reversal of population decline (Table 11.1) and the acknowledgement that Sudbury had one of the most dynamic economies in Canada throughout the latter part of the decade.

Table 11.1 Migration Data for Regional Municipality of Sudbury (from 1980-81 to 1989-90)

Year	In-Migrants	Out-Migrants	Total
1980-81	6, 421	7, 478	-1,057
1981-82	5, 234	6, 299	-1,065
1982-83	4, 193	6, 463	-2, 270
1983-84	4, 373	5, 736	-1, 363

Year	In-Migrants	Out-Migrants	Total
1984-85	4, 230	5, 493	-1, 263
1985-86	4, 601	6, 394	-1, 793
1986-87	5, 519	6, 243	-724
1987-88	6, 149	6, 161	-12
1988-89	7, 224	5, 529	+1, 724
1989-90	7, 599	5, 556	+2,043

Source: Statistics Canada, Migration Estimates from Tax Records for the Post-1981 Period (Ottawa: Administrative Data Development Division).

CONCLUSION

This paper suggests that the history of Sudbury may offer lessons for resource communities victimized by the "staples" trap. While it is true that Sudbury's size and location give it unique advantages, it is nevertheless clear that the "human dynamic" in the form of individual initiative, political leadership, community pride, interaction and entrepreneurial creativity can help direct the path of urban-economic development. The Sudbury experience also indicates that significant change can result from short-term crisis management. In 1982 Sudbury was considered to be a dying community. By the end of the decade, however, it was held up as a national and international example of urban adjustment and revitalization. Sudbury also illustrates the importance of being a metropolis. The attainment of this critical status gave it both economic momentum and enormous political clout. The latter was used successfully to promote state intervention and to diversify the local economy in the health, education, tourism and administrative sectors. With this assistance, Sudbury has managed to attain a much higher degree of self-sustainability. Once known only as a resource community, it has since developed as a major regional service centre for northeastern Ontario.

Notes

[1] Sudbury's early development as a company town of the CPR is traced in Gilbert A. Stelter, "The Origins of a Company Town: Sudbury in the Nineteenth Century," *Laurentian University Review* 3, no.3 (1971): 3-21.

[2] This growth phase can be identified with Thompson's well-known "stage of export specialization." See Wilbur Thompson, *A Preface to Urban Economics* (Baltimore: Johns Hopkins Press, 1965), 15.

[3] International Nickel Company of Canada, *History of Nickel Extraction from Sudbury Ores, Part 1:1846-1920* (Copper Cliff, 1956), 5, 8.

[4] See, for example, "Real Sudbury Please Stand Up," *Globe and Mail, 7* January 1977, 7, and "The New Image of Sudbury – A Diversified Economy and Nice Place to Stay," *Globe and Mail,* 22 March 1977, 10.

[5] Sudbury's incorporation as part of the metropolitan hegemony of Toronto is traced in Gilbert A. Stelter, "Community Development in Toronto's Commercial Empire: The Industrial

Towns of the Nickel Belt, 1883-1931," *Laurentian University Review* 6, no.3 (1974): 3-53.

[6] The early years of Inco's development are traced in John F. Thompson and Norman Beasley, *For the Years to Come: A Story of International Nickel Company of Canada* (Toronto: Longman's, Green & Company, 1960); Robert Stephenson et al., *A Guide to the Golden Age: Mining in Sudbury, 1886-1977* (Sudbury: Laurentian University, Department of History, 1979); O.W. Main, *The Canadian Nickel Industry: A Study in Market Control and Public Policy* (Toronto: University of Toronto Press, 1955); Executive and Operating Staffs, "The Operations and Plants of International Nickel Company of Canada Limited," *Canadian Mining Journal* 67, no.5 (1946); and Jamie Swift and the Development Education Centre, *The Big Nickel: Inco at Home and Abroad* (Kitchener: Between the Lines, 1977), 24.

[7] Gaetan Gervais, "Sudbury, 1883-1914," in *To Our City/A Notre Ville* (Sudbury: Sudbury Centennial Foundation, 1983), 25.

[8] Sudbury's relationship with the northern highway network is outlined in Neil Alexander Wilson, *Evolution of Network Configurations and Optimal Link Additions in the Northern Highway Network* (M.A. thesis, Queen's University, 1975), 33, 111, 116-17.

[9] This theme has been dealt with in H.V. Nelles, *The Politics of Development: Forests, Mines & Hydro-electric Power in Ontario, 1849-1941* (Toronto: Macmillan of Canada, 1974); and "An Alternatives Interview with Elie W. Martel," *Alternatives* 2, no.3 (1973): 10-17.

[10] The development of Falconbridge Nickel Mines Ltd. is outlined in "The Falconbridge Story," *Canadian Mining Journal* 80, no.6 (1959); and John Deverell and the Latin America Working Group, *Falconbridge: Portrait of a Canadian Mining Multinational* (Toronto: James Lorimer & Company, 1975).

[11] Aspects of assessment are dealt with in "Wants Provincial Law Changed to Give Municipalities Right to Tax Mines," *Sudbury Daily Star,* 16 January 1958, 1, 3.

[12] Judith E. Harris, *Weil-Being in Sudbury 1931-1971: A Social Indicator Analysis* (M.Sc. thesis, University of Guelph, 1977), 167.

[13] J.A. Walker, "Planning of Company Towns in Canada," *Canadian Engineer,* 19 July 1927, 147.

[14] Thompson, *A Preface to Urban Economics,* 14-15.

[15] The contention that staple economies frequently stay the same is supported in Mel Watkins, "The Staple Theory Revisited," *Journal of Canadian Studies* 12, no.5 (1977): 86-87; and Pierre L. Bourgeault, *Innovation and Structure*

of Canadian Industry, Science Council of Canada, Special Study No.23 (Ottawa, 1972): 42, 51.

[16] See, for instance, National Security Resources Board, *1950 Materials Survey Nickel* (Washington: United States Department of the Interior, Bureau of Mines, 1952), 1-11.

[17] Swift, *The Big Nickel,* 30; and Deverell, *Falconbridge,* 43-46.

[18] These figures refer to the area now contained by the Regional Municipality of Sudbury.

[19] Charles Dorian, *The First 75 Years: A Headline History of Sudbury, Ontario* (Ilfracombe, England: A.H. Stockwell, 1959), 270.

[20] The concept and importance of the "gateway" is discussed in A. Burghardt, "A Hypothesis about Gateway Cities," *Annals of the Association of American Geographers* 61, no.2 (1971): 269-85.

[21] Richard E. Preston, "The Recent Evolution of Ontario Central Place Systems in the Light of Christaller's Concept of Centrality," *Canadian Geographer* 23, no.3 (1979): 201-21.

[22] The importance of a "main drag" connection is stressed in James and Robert Simmons, *Urban Canada* (Toronto: Copp Clarke, 1969), 66-78.

[23] The functional integration of the regional economy is outlined in George Murphy, *Growth Centre Expansion and Central Place Importance in a Resource-Based Economy* (M.A. thesis,

University of Waterloo, 1979); O.W. Saarinen, "Planning and Other Developmental Influences on the Spatial Organization of Urban Settlement in the Sudbury Area," *Laurentian University Review* 3, no.3 (1971): 44-53; and Ontario Department of Municipal Affairs, *The Sudbury Area: Factors of the Regional Environment* (Toronto, 1960). The existence of political cleavage is acknowledged in J.A. Kennedy, *Sudbury Area Study* (Toronto: Department of Municipal Affairs, 1970).

[24] The early planning history is outlined in O.W. Saarinen, *A Geographical Basis for Regional Planning* (M.A. thesis, University of Western Ontario, 1966).

[25] According to Marshall's typology, Sudbury still exhibited the characteristics of a specialized mining community (John U. Marshall, *The Structure of Urban Systems* (Toronto: University of Toronto Press, 1989), 76-78.

[26] See Richard E. Preston, "The Canadian Central-Place System," in *Cultural Dimensions of Canada's Geography,* edited by F. Helleiner, Peterborough, Trent University, Department of Geography, Occasional Paper No.10 (1984): 296-97 and 307.

[27] Harris, *Weil-Being in Sudbury,* 162.

[28] See, for example, "Boudreau Blames Ignorance for Red Control of Mine-Mill," *Sudbury Daily Star,* 10 December 1959, 1; and "Must

Prove Sudbury Not Communist Area," *Sudbury Star,* 10 October 1959, 13.

[29] See F.E. Emery, "The Next Thirty Years: Concepts, Methods and Anticipations," in *Towards a Social Ecology: Contextual Appreciation of the Future in the Present,* edited by F.E. Emery and E.L. Trist (London: Plenum Press, 1972), 52-56.

[30] "Canadian Mining Town Hits Bottom," *New York Times,* 14 August 1982, 2; and "Sudbury: A City Struggling to Stay Alive," *Toronto Star,* 23 August 1982, Al, A4.

[31] Province of Ontario, *Design for Development Phase Two,* Toronto, Legislature of Ontario, 1968; and Kennedy, *Sudbury Area Study.* Refer also to Ontario Department of Municipal Affairs, *Sudbury: Local Government Reform Proposals* (Toronto, 1971). For a background on regional government see Oiva Saarinen, "Municipal Government in Northern Ontario: An Overview," *Laurentian University Review* 17, no.2 (1985): 5-25.

[32] Regional Municipality of Sudbury, *Official Plan for the Sudbury Area* (Sudbury, 1978).

[33] Val Ross, "The Arrogance of Inco," *Canadian Business* 52, no.5 (May 1979): 44-55, 116-42.

[34] Community Planning Programs Division, Ontario Ministry of Municipal Affairs and Housing, *Project Group on Urban Economic Develop-*

ment Case Study Report: Sudbury, Ontario, Canada (Paris: OECD, 1985), 10.

[35] Hemson Group, *The Hemson Report,* August 1988.

[36] Lesley Barsky, "10 Best Cities to Live in Canada," *Chatelaine,* April 1991, 103; and Peter C. Newman, "Sudbury's Sunny Renaissance," *Maclean's,* 1 April 1991, 40.

[37] Thompson, *A Preface,* 23.

[38] For more details regarding this period, see Oiva Saarinen, "Sudbury: An Historical Case Study of Multiple Urban-Economic Transformation," *Ontario History* 82, no.1 (1990): 53-81.

[39] The importance of urban-based functions to the contemporary economy is affirmed in Peter Hall and Paul Cheshire, "The Key to Success for Cities," *Town & Country Planning,* February 1987, 51.

[40] Wieslaw Michalak, "Economic Changes of the Canadian Urban System 1971-1981," *Alberta Geographer,* no.22 (1986): 65.

[41] A review of the political situation in the Sudbury area during the 1970s is found in Carl Wallace, "Sudbury: The Northern Experiment with Regional Government," *Laurentian University Review* 17, no.2 (1985): 87-101.

[42] Community Planning Programs Division, *Urban Economic Development Case Study.* This study was written by Nigel Richardson. For a history of Sudbury 2001, see *Sudbury 2001, Proceed-*

ings, A Conference on Economic Development: April 6, 7 and 8, 1978 (Sudbury, 1978); and idem, *Retrospect* (Sudbury, 1981). See also "Notice of Intention to Dissolve," *Sudbury Star, 22* January 1987.

[43] "Sudbury Likes New Look," *Spectator,* 10 October 1984, 87. The history of land reclamation in the Sudbury area is outlined in William Lautenbach, *Land Reclamation Program 1978-1984* (Sudbury: Regional Municipality of Sudbury, 1985).

[44] Regional Municipality of Sudbury, *Amendment No.26 to the Official Plan for the Sudbury Planning Area Being the Secondary Plan for the Settlement of Sudbury* (Sudbury, 1987).

[45] Regional Municipality of Sudbury, *Toward Economic Diversification in Sudbury Region* (26 November 1982).

[46] Regional Municipality of Sudbury, *Strategic/Corporate Plan of/by/for the City of Sudbury* (Sudbury, 1987). More recently, environmental technology has been added to the list of proposed development thrusts.

[47] The significance of the white-collar class in reshaping the urban environment is discussed in Kent Gerecke, "New Directions for City Planning," *City Magazine 7,* no.1 (1984: 37); and Richard V. Knight, "City Development in Advanced Industrial Societies," in *Cities in the 21st Century,* edited by Gary Gappert and

Richard V. Knight (Beverly Hills: Sage Publications, 1982), 60.

[48] "Fortunes High, Inco to Hire 160 Full-timers," *Sudbury Star,* 17 March 1988, 1. The changed corporate objectives are outlined in the annual reports of Inco and Falconbridge since 1980.

[49] Joseph E. Schumpeter, *Capitalism, Socialism and Democracy* (New York: Harper & Row, 1962), 81-86. For a similar perspective, see R.D. Norton, "The Once and Present Urban Crisis," *Urban Studies,* 482. For a related perspective, see Margot J. Wojciechowski, "International Trends in Mineral R&D," *CRS Perspectives,* no.29 (November 1988): 3-4.

[50] Thompson, *A Preface,* 15.

[51] Information provided by the Sudbury Regional Development Corporation.

[52] Inco, *Annual Report 1987* (Toronto, 1988), 18.

[53] This development of a strong home base and testing ground as the essential ingredients for a Sudbury-based economic strategy appears to coincide well with the economic strategies currently being advanced by Michael Porter (Patrick Luciani, "Can Canada Stay Rich?" *Canadian Business,* October 1990, 74-77; and David Crane, "U.S. Economist Gets $1 Million to Study Canada," *Toronto Star,* 11 November 1990, Fl-2).

Chapter 12

Fly-In Mining and the Future of the Canadian North

Mark Shrimpton and Keith Storey

Fly-in, or long-distance commuting (LDC), is a form of employment in which the workplace is so isolated from the workers' homes that food and accommodations are provided at the work site and employees spend a fixed number of days working at the site, followed by a fixed number of days at home (Hobart 1979, 2).[1] It is a new variant on the traditional Canadian theme of going away to work in a resource industry. However, it is unique in that it differs from both (1) daily commuting systems, in that workers are away from, and then at home, for extended periods, and (2) seasonal or temporary migration-to-work systems, because of the year-round regularity of the work pattern. The use of fly-in for the exploitation of remote resources in Canada is well established in the oil and gas industry, has grown rapidly in mining and shows signs of increased use in fishing (associated with the use of factory-freezer trawlers), hydro operations and construction projects.

This paper is concerned with the implications of fly-in for the Canadian North. Specifically, the paper describes the increasing use of this system by the Canadian mining industry, discusses the reasons for this, and examines its current and potential implications for the North. Mining-industry data are used to illustrate the themes discussed, but the issues are equally relevant to other types of resource-development activity in northern regions. The themes discussed include employment, economic leakage, out-migration and community impacts. The paper concludes with a discussion of the prospects for the future use of fly-in and how the Canadian North might gain a greater share of the benefits from this system of resource development.

THE USE OF FLY-IN BY THE MINING INDUSTRY

The use of fly-in by the Canadian mining industry has increased rapidly in the last decade. Map 12.1 identifies past and present fly-in operations in Canada and clearly indicates how recent this phenomenon is. The first mine designed to use a fly-in/workcamp system was Asbestos Hill, Quebec, in 1972, but eighteen of the twenty-two operations to date have only been in operation since 1980, and eight have opened since 1987. A total of about 3, 500 miners currently work at fly-in operations. Gold mines (fifteen in total) predominate, and have constituted all new operations since the

Key Lake uranium mine opened in 1983. However, new fly-in copper and uranium mines are being planned and/or developed.

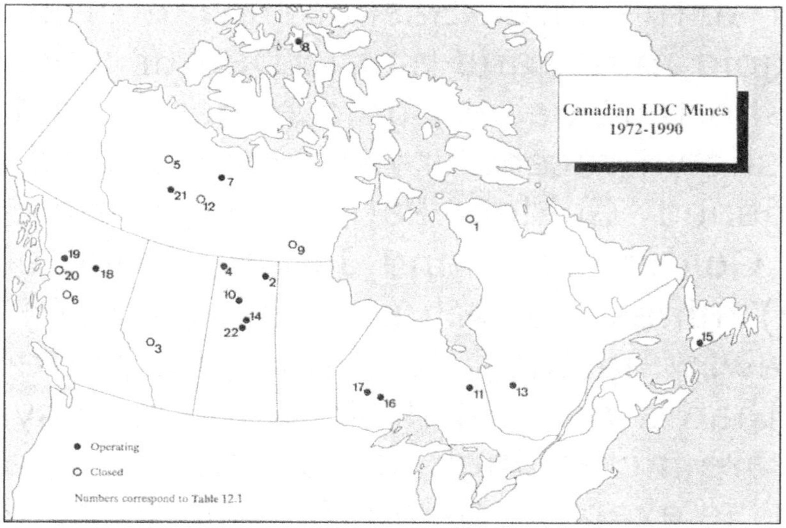

Sources: Mark Shrimpton and Keith Storey.

While fly-in operations constitute only a minority of all Canadian mines, the rate of increase in the use of this system has been rapid. The number of fly-in operations as a percentage of all new gold, uranium and lead/zinc mines increased steadily from 9 percent in 1975-79, to 15 percent in 1980-84, and to 36 percent in 1985-87 (derived from Canada, Department of Energy, Mines and Resources 1987). In some regions, such as the Northwest Territories and northern Saskatchewan, it has become the dominant approach to the development of mineral resources.

Most Canadian fly-in mines use relatively short rotation patterns. The most common one sees employees spending seven days at work followed by seven at home (7/7) (this pattern is used at six of the mines currently in operation, and is worked by approximately 45 percent of the fly-in labour force). It is used at the three Saskatchewan uranium mines and one gold mine, and at the Detour Lake (Ontario) and Lac Shortt (Quebec) gold mines. The 14/14 pattern has shown recent growth, being used at the recently opened Golden Patricia (Ontario), Colomac (N.W.T.), Golden Bear and Johnny Mountain (British Columbia) mines, as well as Lupin (N.W.T.) and Hope Brook (Newfoundland). It is now the pattern worked by more than a third of fly-in workers. Asymmetrical rotations are increasingly uncommon, the largest and most notable example being the Polaris lead/zinc mine in the High Arctic region of the N.W.T.

REASONS FOR THE INCREASING USE OF FLY-IN

The rapid increase in the use of fly-in reflects the hard times recently experienced by the mining industry and the impact these have had on development decisions. Critical factors have included: the limitations of resource towns; the alternative development option; the regulatory and policy environment; transportation and communications technology and infrastructure; and the structure of the mining sector (Storey and Shrimpton 1988). All of these have seen significant changes in recent years, the net result of which strongly favours the fly-in option.

Table 12.1 Characteristics of Current and Former Canadian Mines Using LDC, 1990

Mine	*Deposit*	*Province*	*Opened/Closed*	*Number of LDC Employees*	*Rotation Days in/out*	*Remaining Years of Expected Life*
1. Asbestos Hill	asbestos	Que.	1972-83	c. 400	70/14	-
2. Rabbit Lake	uranium	Sask.	1975	350	7/7	>20
3. Coal Valley	coal	Alta.	1978[1]	335[2]	n.a.	n.a.
4. Cluff Lake	uranium	Sask.	1980	260	7/7	>20
5. Camsell River	silver	N.W.T.	1980-85[3]	200	28/28[4]	-
6. Baker River	gold	B.C.	1981-83	47	14/7[5]	-
7. Lupin	gold	N.W.T.	1982	448	28/14	15-19
					14/14	
8. Polaris	zinc/lead	N.W.T.	1982	275	63/21[6]	15-19
					42/28	
					42/21	
9. Cullaton Lake	gold/silver	N.W.T.	1983-85	130	7/7	-
					42/21	
10. Key Lake	uranium	Sask.	1983	425	7/7	15-19
11. Detour Lake	gold	Ont.	1983	350	7/7[7]	10-14
					4/3	
12. Salmita	gold	N.W.T.	1983-87	25	14/14[8]	-
					35/21	
13. Lac Shortt[9]	gold	Que.	1984	183	14/7	1-3
					7/7	
					5/2	
					4/3	
14. Star Lake	gold	Sask.	1987-89	45	7/7	-
15. Hope Brook[9]	gold	Nfld.	1987	273	14/14	7-9
					4/3	
16. Emerald Lake	gold	Ont.	1988	85[10]	4/4	n.a.
17. Golden Patricia[9]	gold	Ont.	1988	192	14/14	4-6
18. Lawyers Property	gold	B.C.	1988	65	14/14	4
19. Golden Bear[9]	gold	B.C.	1989	111	28/14	4-6
					14/14	
20. Johnny Mountain[9]	gold	B.C.	1989-1990	133	28/14	1-3
					14/14	
21. Colomac[9]	gold	N.W.T.	1990	342	14/14	n.a.
22. Jasper	gold	Sask.	1990	80[11]	7/7	3

1. Workers were bussed in 5 days per week for 8-hour shifts from 1977 to 1978. A daily helicopter-commuting system was used 1978-82 with workers rotating on a 4/4, 12-hour shift system. Since 1982 the company has used a daily bus-commuting system but workers still rotate on the 4/4, 12-hour shift system.
2. 1986 employment which was about 60% capacity.
3. Operations began in 1970 but a rotational system was not adopted until 1980.
4. The original rotation was 54/30 for hourly workers and 36/20 for staff. In 1984 the rotation changed to 28/28 for all employees.

5. 1981 rotation for all personnel was 21/7. This changed in 1982 to 14/7 with the exception of administrators who remained on 21/7.
6. A majority of hourly paid and corporate staff work 63/21, native workers can opt for 42/28, senior management staff work 42/21.
7. Mine staff work 3 weekdays first week, 4 days second week at mine site, except accounting personnel who are located in Timmins.
8. Mill workers worked 14/14, production employees worked 35/21.
9. Data are from 1989 survey. Data for all other mines are from a 1987/1988 survey.
10. Of the total workforce, 55 will commute on a daily basis from River Valley and area, the remaining 30 will live on site during their 4/4 rotation or 4/3 rotation in the case of mill operators.
11. Half are Cameco, and half contractor, employees.

N.B. As of 31 August 1991, operations at Hope Brook, Colomac and Jasper had either been closed or suspended, while the Cominco/Prime Resources Snip mine, a gold/copper operation near Stewart, B.C., came into operation in January 1991.

Source: Prepared from original data collected by Mark Shrimpton and Keith Storey.

LIMITATIONS OF RESOURCE TOWNS

There is a considerable literature on the problems of company or resource towns (see, for example, Robinson 1984; Robson 1988). These problems include:

- lack of economic diversity
- lack of alternative employment opportunities, especially for women
- difficulties of recruiting and retaining top-quality labour
- vulnerability to boom-and-bust cycles
- seasonal instability in terms of employment and income levels
- a limited and often unpredictable life span of the resource
- socio-demographic imbalances
- social problems associated with remote environments, loneliness, alcoholism, etc.
- communities too small to support many urban services, especially in the areas of education, entertainment and retailing
- limitations of the physical, social and political environment
- difficulties of town management
- start-up and wind-down costs to industry and government
- social and economic problems associated with closure

Many of these problems have been exacerbated over time as Canadian miners and their families have come to expect and demand more recreational, educational and other public facilities, as a higher proportion of children seek elementary, secondary and post-secondary education, and as demographic, social and economic changes have made two-income families the norm.

THE REGULATORY AND POLICY ENVIRONMENT

A number of public policy considerations have also affected choices between the resource-town and fly-in alternatives. Principal among these is the fate of single-industry towns when the resource is exhausted or its exploitation becomes uneconomic, thus removing the community's *raison d'etre.* This common Canadian phenomenon has proved costly to the residents of such towns and to the public purse. Accordingly, alternative development options have become more desirable.

At the same time, changes in the regulatory environment have made the construction of mining towns more expensive. The earlier company towns were largely unconstrained by government requirements and controls, leading in many cases to a poor-quality urban environment. More recent mining towns, such as Tumbler Ridge in northeast British Columbia, are, by contrast, conspicuous for the planning effort and investment that have gone into their design. In the case of Tumbler Ridge, the estimated total development cost exceeded $274 million, representing a per capita investment in the order of $45, 700 (McGrath 1986, 232).

Companies contemplating the development of a new town also know they are likely to be subject to an environmental impact-assessment process which may result in costly expenditures before the project is allowed to proceed. Developers are now subject to increasingly onerous regulations regarding urban design, accommodation standards, and servicing, which further add to construction costs.

Another constraint on resource-town development is that increasing federal-provincial involvement in the management of resources has tended to transfer much of the decision making from the private sector to a political-regulatory context. In the mid-1970s, the Alberta government forbade any new town development in the Coal Branch area west of Edmonton. In this case developers had no choice but to use a commuting option (Berg 1986, 31).

A second example of the impact of public policy involves affirmative action programs, particularly as they apply to native peoples. The objective of these programs is to encourage participation in resource development and yet at the same time permit native peoples to maintain elements of their traditional lifestyle. The resource-town strategy does not easily allow these twin objectives to be achieved, whereas fly-in is thought to allow for both the separation of home and workplace and more flexibility in the work schedule. This issue is discussed in more detail later in the paper.

Fly-in employment has also been seen as a means of spreading the economic benefits of resource development to a wide range of communities, native and non-native, rural and urban, and northern and southern. The effect of fly-in is the opposite of that of a resource town, insofar as income generated at the mine is spent in a large number of communities. This contributes to the diversification of the economic bases of these communities and serves to diffuse both the benefits of the mine during its operation and the costs of its closure.

In short, government policy has increasingly encouraged resource developments that address a range of political, economic and social objectives rather than simply economic ones. As a consequence many of the what, where, when and how questions of development must be evaluated against a wider range of variables, the net result being that in many cases, fly-in has become the preferred option.

TRANSPORTATION AND COMMUNICATIONS TECHNOLOGY AND INFRASTRUCTURE

A third element in the new calculus is the reduction in movement and communication costs that has occurred in recent years. New technology, particularly in air transportation and telecommunications, has made the fly-in option far more viable for resource developers than was the case in the 1950s and 1960s. Air transportation is now fast, dependable, relatively safe and inexpensive. Public policy initiatives of the 1960s and 1970s also injected considerable public funds in improving aviation infrastructure. Together, these factors have changed the cost equation in favour of fly-in. Telecommunications links are also very reliable and have declined in absolute cost over time.

THE STRUCTURE OF THE MINING SECTOR

The period from 1981 to 1987 saw a crisis in the mining industry. While the industry has always been vulnerable to short-term mineral cycles and longer-term business cycles, the downturn during this recession coincided with significant structural changes within the industry. This meant that more attention than ever had to be given to improving productivity, rationalizing operations, and reducing production costs so as to ensure industry competitiveness.

One of the main implications for the Canadian metals-mining industry has been the need to increase extraction efficiency, which implies the use of a smaller workforce, a further incentive to fly-in development. Furthermore, recent mineral price fluctuations have caused attention to focus on development options that are more compatible with market variability. The fly-in option presents advantages in that costs of both the start-up and closure of mines, whether temporary or permanent, are reduced. In cases of closure, there are no mining town problems, and the saving in commuter transportation costs is immediate. Fly-in operations are thus both easier and less costly to open and to close (and possibly subsequently to reopen), as is suggested by the closure of the Baker Lake (British Columbia), Cullaton Lake (N.W.T.) and Star Lake (Saskatchewan) mines after only two years of operation.

Fly-in operations also present a number of operational advantages from the company viewpoint. For example, they seem to make it less difficult to attract and retain workers (Nogas 1976, 128; Glass and Lazarovich 1984, 86; Newton 1986, 77). The developers of the Rabbit Lake uranium mine in northern Saskatchewan estimated that if they used a mining town they could expect anything from a 35 to 40 percent annual turnover rate. Turnover started at 28 percent when it opened as a fly-in mine in 1975, and this was down to 5 percent by 1986 (Nogas 1986, 20, 24).

While low turnover rates at most Canadian fly-in operations can, in part, be explained by an absence of alternative mining employment opportunities from 1981 to 1987, there appears, nonetheless, to be a fairly high level of satisfaction with commuting systems. Turnover rates at the Lupin gold mine, for example, have declined since operations began. In 1986 the turnover was in the order of 13 to 14 percent and this declined further in 1987 (Storey and Shrimpton 1989, 77-78).

Operators of fly-in mines have also experienced low levels of absenteeism (Newton 1986, 78), and it has been argued that there are fewer industrial disputes (Rose et al. 1984, 4). However, this last conclusion was based on Australian experience with only four mines. In 1990, there were just four organized fly-in mines in Canada, and these employed 38 percent of the fly-in mine labour force. This contrasts with an overall 50-60 percent level of unionization in the mining sector labour force as a whole (as estimated by the federal Department of Energy, Mines and Resources). To date there have been only two significant disputes at Canadian fly-in mines. One, a short sit-in at Cluff Lake (Saskatchewan), in 1980, was rapidly resolved with the intervention of the Saskatchewan government; the other was a strike by construction workers, not mine workers, over travel conditions to and from the Hope Brook mine in Newfoundland.

FLY-IN AND DEVELOPMENT POLICY

As has been noted above, governments have generally welcomed fly-in because it avoids the use of single-industry towns and their attendant problems, and because it presents the possibility of spreading employment benefits to the native and non-native residents of small northern communities. Similarly, when the mine closes and jobs are lost, the impact is not limited to one or two communities in the immediate area of the mine, but is spread over many centres.

Other indirect economic benefits for northern residents result from the tax revenues governments receive from fly-in mines. Furthermore, fly-in permits individuals to live in the communities of their choice where, as consumers and taxpayers, they may contribute significantly to the continued viability of those communities. Finally, where resource developments are strictly controlled short-term phenomena, the net effect of fly-in operations may be to minimize the environmental damage to sensitive, natural settings.

However, there has been increasing recognition that fly-in may have negative effects. One of these is the potential for "fly-over," whereby the labour force and communities close to such mines gain few benefits because the mines draw their labour and other requirements from larger, more distant centres. These workers, supplies and services may come not only from distant parts of the region, but from entirely different territories, provinces and states.

Another concern results from the fact that, because in most cases the company pays for travel to and from designated "pick-up" points, no significant transportation costs accrue to individual workers. They are free to move from a small community to a larger one with a greater range of services. Alternatively, they may make the reverse move so as to reduce housing costs and live in a community felt to be less stressful and more suitable for raising a family.[2] If the latter alternative is selected, as it might well be by younger workers, demographic imbalances may result.

FLY-OVER OF EMPLOYMENT

While fly-in mining provides considerable scope for distributing employment benefits throughout northern regions, it also has resulted in a large amount of employment fly-over. Most of the direct and spin-off employment benefits from a conventional mine are retained in the North, but there is considerable leakage to the south in the case of fly-in.

Table 12.2 summarizes data on the places of residence of workers at eleven fly-in mines. Clearly the length of rotation and the transportation policies of companies significantly affect the regional distribution. At one extreme, the Detour Lake (Ontario) gold mine mostly uses a 7/7 day rotation and only flies workers to and from two Northern Ontario mining communities, Cochrane and Timmins. Virtually all of its employees live in the region, with only 4.5 percent coming from southern Ontario and 1.5 percent from other provinces.

The British Columbian, Saskatchewan and Newfoundland fly-in mines also use relatively short rotations (7/7 and 14/14), only have intra-provincial pick-up points and exhibit a considerable concentration of employment. However, as exemplified by the Saskatchewan mines, there may still be considerable fly-over of northern communities. This latter group of mines have both northern and southern pickup points, and in the case of both the Key Lake and Rabbit Lake mines, about 80 percent of the labour force lives in the south, mostly around Saskatoon and Prince Albert. Cluff Lake, which has had a very effective native-hiring program, employs a much higher proportion (approximately 50 percent) of northern residents.

Table 12.4
Purchasing Profiles, Canadian LDC Mines, 1989
(percent values)

	Mine					
Region	*Johnny Mountain B.C.*	*Golden Bear B.C.*	*Golden Patricia Ont.*	*Detour Lake Ont.*	*Lac Short Que.*	*Hope Brook Nfld.*
N.W.T.-Yellowknife	-	-	-	-	-	-
N.W.T.-Other	-	-	-	-	-	-
Yukon	-	20	-	-	-	-
British Columbia	65	55	-	-	-	1
Alberta-Northern	-	-	-	-	-	-
Alberta-Southern	-	-	-	-	-	5
Saskatchewan-Northern	-	-	-	-	-	-
Saskatchewan-Southern	-	-	-	-	-	-
Manitoba-Northern	-	-	-	-	-	-
Manitoba-Southern	-	-	9	-	-	-
Ontario-Northern	-	-	88	80	13	-
Ontario-Southern	-	-	3	15	13	33
Quebec-Northern	-	-	-	5	73	5
Quebec-Southern	-	-	-	5	-	-
Maritimes	-	-	-	-	-	10
Newfoundland	-	-	-	-	-	47
U.S.A.	35[1]	25	-	-	-	-

1. 1987 data. 2. 1989 data. 3. Does not include catering staff.
Source: Prepared from original data collected by Mark Shrimpton and Keith Storey.

The Northwest Territories fly-in mines also have pick-up points in both the N.W.T. and the south. Lupin provides, at no expense to the workers, jet service to and from Edmonton and Yellowknife, and turboprop services to other pick-up points in the N.W.T. Polaris, which has a very long asymmetrical rotation (mostly 63/21), provides a free link to Resolute and pays for commercial air transportation between there and any airport in Canada. In the case of both Lupin and Polaris, workers live in all the main economic regions of the country. In the case of Polaris, there are also employees who live in the Caribbean, Venezuela and United Kingdom. In total, only 10.4 percent of the employees at these N.W.T. fly-in mines live in the Territories.

The amount of employment leakage from a region also depends on the location of the mine within it. For example, the Golden Patricia mine in northwestern Ontario, only has pick-up points in that region (Dryden, Sioux Lookout, Pickle Lake, Cat Lake and Slate Falls). However, because of the proximity of the Manitoba border and Winnipeg, the distance to urban centres in Ontario, and the 14/14 rotation, half of the labour force lives outside the province. Thirty percent of all workers live in Manitoba, while 8 percent commute as far as the Yukon and British Columbia.

Many of those who choose to live in the south have their homes in major urban centres. A survey of workers at the Key Lake uranium mine in northern Saskatchewan showed that 57 percent lived in Saskatoon, Regina, Winnipeg or "a city." Of the respondents to a survey of Polaris miners, 55 percent lived in major metropolitan areas: Vancouver, Calgary,

Edmonton, Winnipeg, Sudbury, St. Catharines, Toronto, Ottawa, Montreal or Halifax.

The commuting of workers to and from the south clearly has both direct and indirect economic effects. Not only do these miners spend virtually all their payroll in the south, generating income and employment multipliers, but also they are taxed in a different municipal and, frequently, provincial/territorial jurisdiction. It is also in the south that their families will live, spend any independent income (evidence from Key Lake and Polaris shows that 40 to 45 percent of miners' spouses are in paid employment), and will be taxed, although these family members are also likely to demand significant social, health and educational services there.

Table 12.3 Purchasing Profiles, LDC and Non-LDC Mines: Northwest Territories, Canada, 1986

		LDC Mines[1] (N=2)	*Non-LDC Mines*[2] (N=3)	*Total*
(N=5)				
Total Purchases:		65,219,000	28,271,000	93,490,000
Source (percent):				
	Northwest Territories	19.3	34.7	23.7
	Alberta/British Columbia	30.0	2.8	37.0
	Saskatchewan/Manitoba	1.5	3.4	2.3
	Ontario/Quebec	43.0	8.0	32.3
	United States	4.1	1.0	3.2
	Other	2.3	0.0	1.5
Total		100.0	100.0	100.0

Source: Derived from Stephenson Kellogg Ernst and Whinney, 1987, Exhibit III-3, p.8.

1. Lupin, Nanisivik, Polaris.

2. Giant Yellowknife, Nerco Con, Tom.

N.B. Nanisivik mine, which is a hybrid, using a mixture of LDC and daily commute from Arctic Bay, is excluded from this analysis.

Fly-over Of Business Opportunities

Table 12.4
Purchasing Profiles, Canadian LDC Mines, 1989
(percent values)

	Mine					
Region	*Johnny Mountain B.C.*	*Golden Bear B.C.*	*Golden Patricia Ont.*	*Detour Lake Ont.*	*Lac Short Que.*	*Hope Brook Nfld.*
N.W.T.-Yellowknife	-	-	-	-	-	-
N.W.T.-Other	-	-	-	-	-	-
Yukon	-	20	-	-	-	-
British Columbia	65	55	-	-	-	1
Alberta-Northern	-	-	-	-	-	-
Alberta-Southern	-	-	-	-	-	5
Saskatchewan-Northern	-	-	-	-	-	-
Saskatchewan-Southern	-	-	-	-	-	-
Manitoba-Northern	-	-	-	-	-	-
Manitoba-Southern	-	-	9	-	-	-
Ontario-Northern	-	-	88	80	13	-
Ontario-Southern	-	-	3	15	13	33
Quebec-Northern	-	-	-	5	73	5
Quebec-Southern	-	-	-	5	-	-
Maritimes	-	-	-	-	-	10
Newfoundland	-	-	-	-	-	47
U.S.A.	35[1]	25	-	-	-	-

Source: K. Storey and M. Shrimpton, unpublished survey of Canadian LDC mines, 1989. 1. Alaska.

Just as there is a fly-over of employment, there is a fly-over of local businesses whenever supplies and services are acquired in the south. As is the case with employment patterns, this is most clearly the case in the N.W.T. Data contrasting the purchasing patterns of fly-in and conventional mines there (Table 12.3) show that while the N.W.T.'s three conventional mines make 34.7 percent of their total expenditures on supplies and services within the territory, the two fly-in mines only spend 19.3 percent there. They also buy less in British Columbia and the Prairies (31.5 percent, as against 56.2 percent for conventional mines). Instead, the fly-in mines are heavily dependent on suppliers in Ontario and Quebec (43 percent, as against 8 percent for the conventional mines) and outside Canada (6.4 percent as against 1.0 percent).

However, more recent data for a selection of fly-in mines in British Columbia, Ontario, Quebec and Newfoundland show much less flyover (Table 12.4). In all cases except the Hope Brook gold mine, more than half of all supplies and services are purchased from the region in which the mine is located. The two Ontario mines, Golden Patricia and Detour Lake, have the highest proportion of local purchases, acquiring 88 percent and 80 percent respectively from Northern Ontario. While comparative figures are not available, it seems unlikely that these proportions are significantly less than those for conventional mines located in the region.

A key factor here is clearly the existence of specialist suppliers at a convenient location within the region where the mine is located. Timmins, Sudbury and Thunder Bay, for example, are easily able to meet the needs of the Northern Ontario fly-in mines, which are both connected to the provincial highway network. Golden Bear is near the Yukon border and has good transportation links with it; however, the reasons for the high degree of reliance of the B.C. mines on the

United States is not known. At the other extreme, the island of Newfoundland (as against Western Labrador), has a relatively poorly developed mining-services industry. The N.W.T. has a strong mining-industry service sector, but it is concentrated in Yellowknife, which has poor links with Polaris. The mine ships most of its ore direct to Europe by sea, and much equipment is brought in on the same vessels. Other supplies come by sea from Montreal, while the main air links are to Edmonton (via Yellowknife) and Montreal (via Iqualuit). This limits the ability of Yellowknife businesses to supply the mine, as is reflected by the fact that only 7 percent of all Polaris purchases are made within the N.W.T.

The Lupin Mine has more direct air and ground links to Yellowknife, the latter through a seasonal winter road. However, it still only makes 23 percent of purchases in the N.W.T., compared to the 34.7 percent average for the conventional mines which are located near Yellowknife.

MIGRATION

The fly-over of employment and, to a lesser extent, supplies and services means that the contribution fly-in mines make to the North is more modest than might otherwise be expected. Indeed, fly-in operations may have a negative demographic, economic and social effect on the North because they remove some of its more skilled and motivated residents. Many companies and governments have actively encouraged the hiring of native and non-native northerners, but when they find themselves working alongside employees who live and have families in the south, significant numbers of those hired in the North choose to relocate there.

For example, human resources staff at Polaris have reported that

> as of June 1987 the current number of employees hired from the north was 71, however with the years 38, (54 percent) of these have moved south. The main reason attributed to the desire for a change of place of residence, is to allow a family to be closer to friends and family to provide support during the lengthy periods of absence of the spouse. (Storey and Shrimpton 1989, 43)

Lupin has also seen some decline in the numbers of workers living in Yellowknife "as people choose to move south for cheaper living conditions," while a Saskatchewan uranium mine reported that "a number of northern employees have moved to the larger centres such as Saskatoon." Surveys of Polaris and Key Lake workers confirm this, with 42 percent and 17 percent of the respondents respectively stating that they have moved their homes because of the fly-in work system. While no question specifically sought data on the places they moved from and to, many provided such information. In the case of Key Lake, a majority had moved to be closer to the pick-up points (of which Saskatoon and Prince Albert are the most commonly used), while two of twenty-two respondents who had moved indicated that they had moved south to get away from "isolated" northern communities. Overall, 74 percent of all movers stated that their current place of residence was Saskatoon, indicating that one in ten of all Key Lake employees had moved to that city because of the work pattern.

Courtesy of Keith Storey

Accommodations at Cominco's Polaris zinc/lead operation, Polaris Mine, Little Cornwallis Island, Northwest Territories.

A move closer to a pick-up point is more likely to be important to Key Lake workers, where the 7/7 schedule requires a round-trip commute every two weeks, than at Polaris, with its 63/21 rotation and associated round trip every three months. Only one of thirteen Polaris respondents cited a move to a pick-up point as the reason for changing the place of residence, and he saw this as secondary to moving to "a warmer climate and larger city." In total, seven (54 percent) of the movers said they moved to live in a better climate or a "nice place for R & R," while six (46 percent) mentioned a desire to be near their, or their spouse's, family: "so my wife can be closer to parents and sister," and so on. Other reasons given included "good schooling for wife and kids" and "Yellowknife – too cold, too expensive."

Courtesy of Keith Storey

Accommodations at Cameco's Key Lake uranium operation, Key Lake, Saskatchewan.

Those Polaris movers who indicated where they had moved from had lived in the N.W.T. (five respondents), Ontario (three), British Columbia (two), and Yukon, Alberta and Saskatchewan (one each). They had moved to British Columbia (seven), Alberta (two), Ontario (two) and England (one), while one no longer had a fixed residence. There was also a general move from small to large communities, with two workers mentioning this as a factor in their decision to move. Specific destinations included Vancouver, Calgary and

St. Catharines. Another factor which may have fuelled this move was the greater availability of employment opportunities for spouses (with two-income couples now being the norm among Canadian families) and children (Shrimpton and Storey 1989).

As with overall employment fly-over, the scope for this phenomenon of out-migration is clearly dependent on both the work schedule and the range and location of pick-up points. The situation at Detour Lake, with a 7/7 pattern and only Northern Ontario pick-up points, presents little potential for out-migration, while Polaris is at the other extreme. Generally speaking, both fly-over and out-migration will be reduced as the rotation becomes shorter and the pick-up points more localized.

IMPLICATIONS FOR THE FUTURE OF THE CANADIAN NORTH

The above analysis suggests that, in the absence of significant economic or policy change, fly-in will have major effects on patterns of development in the Canadian North. These include the following:

No New Mining Towns Trends in mineral economics, technology, work schedules and government policy all strongly, and increasingly, favour fly-in over the mining-town option. No new mining town has been built in the last decade, while, during the same period, sixteen fly-in operations have opened. It is not clear under what circumstances construction of a new mining town would now be deemed feasible and/or desirable.

Closure of Existing Mining Towns As resource depletion or market forces make existing operations uneconomic and in the absence of other employment and business opportunities, mining towns will close.[3] This is a long-standing Canadian phenomenon. What is different is that these communities are not being replaced by new mining towns elsewhere.

Increasing Concentration of Population The population of the North will be increasingly concentrated in either the larger administrative and service centres, such as Yellowknife, Thompson, Sudbury and Timmins, or in smaller, predominantly native, communities. Some of the residents of both may work at fly-in mines, as in the case of the Yellowknife residents at Polaris and Lupin, the Timmins miners who work at Detour Lake, and the workers at both Saskatchewan and N.W.T. mines who are flown in from small northern villages.

Fly-Over and Out-Migration The above evidence suggests that, as conventional mines continue to close and new fly-in mines open, the leakage of mining employment and business from the North associated with the latter development option will increase. Similarly, it seems likely that there will be a continued out-migration of those hired from the North by the operators of existing and new fly-in mines. Climate, cost of living, family and other lifestyle factors will continue to promote such movement, as will the fact that the fly-in commute schedule allows workers to enjoy the benefits of living in major metropolitan areas while avoiding the congestion peaks and some other disadvantages of living there.

Clearly, fly-in has the potential to cause major changes in the Canadian North, especially given the fact that it is also increasingly being adopted by the hydro, oil and gas, and construction industries. It is not unreasonable to expect similar consequences in the case of those industries.

It will be difficult for governments to know whether, and/or how, to respond to these trends. The decision as to whether to respond is dependent on fundamental attitudes towards the North and development – for example, the vexed question of who should be the beneficiaries of northern development. The question of how to respond is plagued by uncertainty about what carrots or sticks will be effective in influencing these trends. The experience of fly-in to date, however, does suggest its potentials as a tool for shaping the North. For example:

Northern Benefits It may be possible to reduce fly-over and out-migration by the inclusion of greater northern benefits provisions in the terms of development agreements. Northern and/or native hiring requirements are already common in Saskatchewan and, to a lesser extent, the N.W.T. The use of northern commuting networks, limited access to southern pick-up points, and shorter rotations would have some effect of limiting fly-over and out-migration. However, they will also increase industry costs and companies may have problems finding an adequate skilled northern labour force.

Native Hiring Fly-in mines have considerable potential as sources of native employment. Given an appropriate rotation pattern, hiring and training policies, workcamp culture and so on, fly-in may be one of the best ways of allowing native northerners to combine industrial employment with a traditional lifestyle (see Hobart 1989).

The compatibility of the work pattern with native lifestyles may be a limiting factor in the numbers of fly-in miners drawn from small northern communities. The high levels of native employment at the Cluff Lake uranium mine suggest the potential of fly-in, and it will be interesting to see how successful the Red Dog (Alaska) mine will be in this regard. The work pattern and policies there have been developed jointly by Cominco and the Red Dog Band Council.

Commuting Centres In western Newfoundland, fly-in has permitted some workers to remain in conventional mining towns after the mine has closed, commuting to new fly-in operations. There is no reason why a mining town such as Elliot Lake, which has both a pool of unemployed skilled labour and a well-equipped airport, should not become a pick-up point for distant mines.

CONCLUSION

Fly-in is increasingly used by the mining and other resource industries, especially for operations in the North and peripheral regions. The principal advantages of the system from a public policy perspective are that it makes the creation of single-industry towns, with their attendant problems, unnecessary, and it allows the economic benefits of the activity (and the costs of closures) to be widely spread.

The above examination of patterns of employment and expenditure by fly-in mines indicates that they may cause a leakage of many benefits, including skilled labour, to the south. This has occurred despite some government and industry initiatives seeking to channel these benefits to native and non-native northern populations. In some cases, such as the out-migration of workers hired in the North, these initiatives have proved counter-productive.

Despite this, there is evidence that fly-in has potential as a tool for northern development. This potential will only be fulfilled if care is taken to ensure that fly-in systems are appropriate to the needs of northern residents, both in facilitating participation in the benefits of development, and in ensuring that it complements existing socio-economic structures and relationships. If a satisfactory compromise is to be reached among the competing goals of minimizing industry development and operating costs, minimizing the social and economic costs of non-renewable resource projects, maximizing northern benefits, and achieving other northern development objectives, industry, governments and northern communities will have to adopt a much more sophisticated approach to the use of fly-in.

Notes

[1] While the most common method of commuting to and from these mines is by air and the term "fly-in" is widespread in the industry, there are operations which use road and sea transportation. Long-distance commuting (LDC) is an alternative label coined by Hobart, although even it may not be appropriate in the case of some recent mines which use a relatively short commute. Alternative labels such as "commuter" and "residential" mining have been proposed. The terms fly-in, long-distance commuting and LDC are used interchangeably in this paper.

[2] Even where the company pays only the transportation costs from specific pick-up points, workers may still find it cheaper or otherwise preferable to live at the place of their choice and pay the additional transportation costs themselves.

[3] Some such communities may be able to survive by providing workers to nearby new mining operations on a daily commute basis, but while highway improvements may increase the possi-

bility of this happening, the growing use of extended workdays predicates against it by reducing the feasible duration of the daily journey to and from work.

References

Berg, L. 1986. "From Towns to Wings to Wheels at Stereo." In *Towns, Wheels or Wings for Resource Development?,* 30-32. Selected conference papers from the 3-4 February 1986 conference, Vancouver, B.C., organized by the Institute for Research on Public Policy (IRPP), Victoria, B.C.

Canada. Department of Energy, Mines and Resources. 1987. "Current and Past Producers in *Canada,* 07/15/87." Unpublished data, National Mineral Inventory.

Glass, R.D., and J. Lazarovich. 1984. "The Government Perspective." In *Mining Communities: Hard Lessons for the Future.* Proceedings, No.14, Centre for Resource Studies, Queen's University, Kingston, Ontario.

Hobart, C.W. 1979. "Commuting Work in the Canadian North: Some Effects on Native People." *Proceedings; Conference on Commuting and Northern Development,* 1-38. University of Saskatchewan, Institute of Northern Studies, Saskatoon, 15, 16 February.

_____. 1989. "Company Town or Commuting: Implications for Native People." In *The Commuting Alternative: A Contemporary Response to Community Needs in the Resource Sector,* edited by R. Robson. Winnipeg: Institute of Urban Studies, University of Winnipeg.

McGrath, S. 1986. "Tumbler Ridge: An Assessment of the Local Government Method of Resource Community Development." *Impact Assessment Bulletin* 4, no.1-2: 211-36.

Newton, P.W. 1986. "Settlement Options for Resource Development in Australia." In *Towns, Wheels or Wings for Resource Development?,* 54-87. Selected conference papers from the 3-4 February 1986 conference, Vancouver, B.C., organized by IRPP, Victoria, B.C.

Nogas, F.R. 1986. "Towns Versus Commuting at the Rabbit Lake Uranium Mine." In *Towns, Wheels or Wings for Resource Development?,* 20-24. Selected conference papers from the 3-4 February 1986 conference, Vancouver, B.C., organized by IRPP, Victoria, B.C.

Robinson, I.M. 1984. "New Resource Towns on Canada's Frontier: Selected Contemporary Issues." In D.D. Detomasi and J.W. Gartrell, *Resource*

Communities: A Decade of Disruption (Boulder, Col.: Westview Press), 1-21.

Robson, R. 1988. "The Decline of Resource Towns." Paper presented to the Canadian Urban and Housing Studies Conference, Winnipeg, Manitoba, 18-20 February.

Rose, D., et al. 1984. "Argyle Project – The Case for Commuting." In *Proceedings of the Australasian Institute of Mining and Metallurgy Conference.* Darwin, August, 1-8.

Shrimpton, M., and K. Storey. 1989. "The Urban Miner: Long Distance Commuting to Work in the Mining Sector and Its Implications for the Canadian North." In 77k *Commuting Alternative: A Contemporary Response to Community Needs in the Resource Sector,* edited by R. Robson. Winnipeg: Institute of Urban Studies.

Stevenson Kellogg Ernst & Whinney. 1987. *Strategies to Improve Northwest Territories Business Opportunities Related to Mining and Exploration.* Report prepared for the Energy, Mines and Resources Secretariat, Government of the Northwest Territories.

Storey, K., and M. Shrimpton. 1988. *Long Distance Commuting in the Canadian Mining Industry.* Working

Paper No.43, Centre for Resource Studies, Queen's University, Kingston, Ontario.

_____. 1989. *Impacts on Labour of Long Distance Labour Commuting in the Canadian Mining Industry.* ISER Report No.3, Institute for Social and Economic Research, Memorial University of Newfoundland, St. John's, Newfound-land.

Made in the USA
Monee, IL
05 November 2026

67979970R00201